CAMBRIDGE LIBRARY (

Books of enduring scholarly

Philosophy

This series contains both philosophical texts and critical essays about philosophy, concentrating especially on works originally published in the eighteenth and nineteenth centuries. It covers a broad range of topics including ethics, logic, metaphysics, aesthetics, utilitarianism, positivism, scientific method and political thought. It also includes biographies and accounts of the history of philosophy, as well as collections of papers by leading figures. In addition to this series, primary texts by ancient philosophers, and works with particular relevance to philosophy of science, politics or theology, may be found elsewhere in the Cambridge Library Collection.

Collected Essays

Known as 'Darwin's Bulldog', the biologist Thomas Henry Huxley (1825–95) was a tireless supporter of the evolutionary theories of his friend Charles Darwin. Huxley also made his own significant scientific contributions, and he was influential in the development of science education despite having had only two years of formal schooling. He established his scientific reputation through experiments on aquatic life carried out during a voyage to Australia while working as an assistant surgeon in the Royal Navy; ultimately he became President of the Royal Society (1883–5). Throughout his life Huxley struggled with issues of faith, and he coined the term 'agnostic' to describe his beliefs. This nine-volume collection of Huxley's essays, which he edited and published in 1893–4, demonstrates the wide range of his intellectual interests. Volume 3 contains lectures and essays spanning thirty years of campaigning about the importance of science in education.

Cambridge University Press has long been a pioneer in the reissuing of out-of-print titles from its own backlist, producing digital reprints of books that are still sought after by scholars and students but could not be reprinted economically using traditional technology. The Cambridge Library Collection extends this activity to a wider range of books which are still of importance to researchers and professionals, either for the source material they contain, or as landmarks in the history of their academic discipline.

Drawing from the world-renowned collections in the Cambridge University Library, and guided by the advice of experts in each subject area, Cambridge University Press is using state-of-the-art scanning machines in its own Printing House to capture the content of each book selected for inclusion. The files are processed to give a consistently clear, crisp image, and the books finished to the high quality standard for which the Press is recognised around the world. The latest print-on-demand technology ensures that the books will remain available indefinitely, and that orders for single or multiple copies can quickly be supplied.

The Cambridge Library Collection will bring back to life books of enduring scholarly value (including out-of-copyright works originally issued by other publishers) across a wide range of disciplines in the humanities and social sciences and in science and technology.

Collected Essays

VOLUME 3: SCIENCE AND EDUCATION

THOMAS HENRY HUXLEY

CAMBRIDGE
UNIVERSITY PRESS

CAMBRIDGE UNIVERSITY PRESS

Cambridge, New York, Melbourne, Madrid, Cape Town,
Singapore, São Paolo, Delhi, Tokyo, Mexico City

Published in the United States of America by Cambridge University Press, New York

www.cambridge.org
Information on this title: www.cambridge.org/9781108040532

This edition first published 1893
This digitally printed version 2011

ISBN 978-1-108-04053-2 Paperback

COLLECTED ESSAYS

By T. H. HUXLEY

VOLUME III

SCIENCE & EDUCATION

ESSAYS

BY

THOMAS H. HUXLEY

London
MACMILLAN AND CO.
1893

RICHARD CLAY AND SONS, LIMITED,
LONDON AND BUNGAY.

PREFACE

THE apology offered in the Preface to the first volume of this series for the occurrence of repetitions, is even more needful here I am afraid. But it could hardly be otherwise with speeches and essays, on the same topic, addressed at intervals, during more than thirty years, to widely distant and different hearers and readers. The oldest piece, that "On the Educational Value of the Natural History Sciences," contains some crudities, which I repudiated when the lecture was first reprinted, more than twenty years ago; but it will be seen that much of what I have had to say, later on in life, is merely a development of the propositions enunciated in this early and sadly-imperfect piece of work.

In view of the recent attempt to disturb the compromise about the teaching of dogmatic theology, selemnly agreed to by the first School Board for London, the fifteenth Essay; and, more particularly, the note on p. 388, may be found interesting.

T. H. H.

HODESLEA, EASTBOURNE,
September 4th, 1893.

CONTENTS

viii

V

VI

VII

VIII

IX

X

COLLECTED ESSAYS

VOLUME III

I

JOSEPH PRIESTLEY

[1874]

If the man to perpetuate whose memory we have this day raised a statue had been asked on what part of his busy life's work he set the highest value, he would undoubtedly have pointed to his voluminous contributions to theology. In season and out of season, he was the steadfast champion of that hypothesis respecting the Divine nature which is termed Unitarianism by its friends and Socinianism by its foes. Regardless of odds, he was ready to do battle with all comers in that cause ; and if no adversaries entered the lists, he would sally forth to seek them.

To this, his highest ideal of duty, Joseph Priestley sacrificed the vulgar prizes of life, which, assuredly, were within easy reach of a man of his singular energy and varied abilities. For this object he put aside, as of secondary importance, those scientific investigations which he loved so well, and in which he showed himself so competent to enlarge the boundaries of natural knowledge and to win fame. In this cause he not only cheerfully suffered obloquy from the bigoted and the unthinking, and came within sight of martyrdom; but bore with that which is much harder to be borne than all these, the unfeigned astonishment and hardly disguised contempt of a brilliant society, composed of men whose sympathy and esteem must have been most dear to him, and to whom it was simply incomprehensible that a philosopher should seriously occupy himself with any form of Christianity.

It appears to me that the man who, setting before himself such an ideal of life, acted up to it consistently, is worthy of the deepest respect, whatever opinion may be entertained as to the real value of the tenets which he so zealously propagated and defended.

But I am sure that I speak not only for myself, but for all this assemblage, when I say that our purpose to-day is to do honour, not to Priestley, the Unitarian divine, but to Priestley, the fearless defender of rational freedom in thought and in action: to Priestley, the philosophic thinker; to that Priestley who held a foremost place among "the swift runners who hand over the lamp of life,"[1] and transmit from one generation to another

---

[1] "Quasi cursores, vitaï lampada tradunt." — LUCR. *De Rerum Nat.* ii. 78.

the fire kindled, in the childhood of the world, at the Promethean altar of Science.

The main incidents of Priestley's life are so well known that I need dwell upon them at no great length.

Born in 1733, at Fieldhead, near Leeds, and brought up among Calvinists of the straitest orthodoxy, the boy's striking natural ability led to his being devoted to the profession of a minister of religion; and, in 1752, he was sent to the Dissenting Academy at Daventry—an institution which authority left undisturbed, though its existence contravened the law. The teachers under whose instruction and influence the young man came at Daventry, carried out to the letter the injunction to "try all things : hold fast that which is good," and encouraged the discussion of every imaginable proposition with complete freedom, the leading professors taking opposite sides ; a discipline which, admirable as it may be from a purely scientific point of view, would seem to be calculated to make acute, rather than sound, divines. Priestley tells us, in his "Autobiography," that he generally found himself on the unorthodox side : and, as he grew older, and his faculties attained their maturity, this native tendency towards heterodoxy grew with his growth and strengthened with his strength. He passed from Calvinism to Arianism ; and finally, in middle life,

landed in that very broad form of Unitarianism
by which his craving after a credible and consistent theory of things was satisfied.

On leaving Daventry Priestley became minister
of a congregation, first at Needham Market, and
secondly at Nantwich ; but whether on account of
his heterodox opinions, or of the stuttering which
impeded his expression of them in the pulpit, little
success attended his efforts in this capacity.   In
1761, a career much more suited to his abilities
became open to him.   He was appointed " tutor
in the languages " in the Dissenting Academy at
Warrington, in which capacity, besides giving three
courses of lectures, he taught Latin, Greek, French,
and Italian, and read lectures on the theory of
language and universal grammar, on oratory,
philosophical criticism, and civil law.   And it is
interesting to observe that, as a teacher, he encouraged and cherished in those whom he instructed the freedom which he had enjoyed, in his
own student days, at Daventry.   One of his pupils
tells us that,

" At the conclusion of his lecture, he always encouraged his
students to express their sentiments relative to the subject of it,
and to urge any objections to what he had delivered, without
reserve.   It pleased him when any one commenced such a conversation.   In order to excite the freest discussion, he occasionally
invited the students to drink tea with him, in order to canvass
the subjects of his lectures.   I do not recollect that he ever
showed the least displeasure at the strongest objections that
were made to what he delivered, but I distinctly remember the

smile of approbation with which he usually received them : nor did he fail to point out, in a very encouraging manner, the ingenuity or force of any remarks that were made, when they merited these characters. His object, as well as Dr. Aikin's, was to engage the students to examine and decide for themselves, uninfluenced by the sentiments of any other persons." [1]

It would be difficult to give a better description of a model teacher than that conveyed in these words.

From his earliest days, Priestley had shown a strong bent towards the study of nature; and his brother Timothy tells us that the boy put spiders into bottles, to see how long they would live in the same air—a curious anticipation of the investigations of his later years.   At Nantwich, where he set up a school, Priestley informs us that he bought an air pump, an electrical machine, and other instruments, in the use of which he instructed his scholars.   But he does not seem to have devoted himself seriously to physical science until 1766, when he had the great good fortune to meet Benjamin Franklin, whose friendship he ever afterwards enjoyed.   Encouraged by Franklin, he wrote a " History of Electricity," which was published in 1767, and appears to have met with considerable success.

In the same year, Priestley left Warrington to become the minister of a congregation at Leeds ;

[1] *Life and Correspondence of Dr. Priestley*, by J. T. Rutt. Vol. I. p. 50.

and, here, happening to live next door to a public
brewery, as he says,

"I, at first, amused myself with making experiments on the
fixed air which I found ready-made in the process of fermentation. When I removed from that house I was under the
necessity of making fixed air for myself; and one experiment
leading to another, as I have distinctly and faithfully noted in
my various publications on the subject, I by degrees contrived
a convenient apparatus for the purpose, but of the cheapest
kind.

"When I began these experiments I knew very little of
chemistry, and had, in a manner, no idea on the subject before
I attended a course of chemical lectures, delivered in the
Academy at Warrington, by Dr. Turner of Liverpool. But I
have often thought that, upon the whole, this circumstance
was no disadvantage to me; as, in this situation, I was led
to devise an apparatus and processes of my own, adapted
to my peculiar views; whereas, if I had been previously
accustomed to the usual chemical processes, I should not
have so easily thought of any other, and without new modes of
operation, I should hardly have discovered anything materially
new."[1]

The first outcome of Priestley's chemical work,
published in 1772, was of a very practical character. He discovered the way of impregnating
water with an excess of "fixed air," or carbonic
acid, and thereby producing what we now know
as "soda water"—a service to naturally, and
still more to artificially, thirsty souls, which those
whose parched throats and hot heads are cooled
by morning draughts of that beverage, cannot
too gratefully acknowledge. In the same year,
Priestley communicated the extensive series of

[1] *Autobiography*, §§ 100, 101.

observations which his industry and ingenuity had accumulated, in the course of four years, to the Royal Society, under the title of " Observations on Different Kinds of Air "—a memoir which was justly regarded of so much merit and importance, that the Society at once conferred upon the author the highest distinction in their power, by awarding him the Copley Medal.

In 1771 a proposal was made to Priestley to accompany Captain Cook in his second voyage to the South Seas. He accepted it, and his congregation agreed to pay an assistant to supply his place during his absence. But the appointment lay in the hands of the Board of Longitude, of which certain clergymen were members ; and whether these worthy ecclesiastics feared that Priestley's presence among the ship's company might expose His Majesty's sloop *Resolution* to the fate which aforetime befell a certain ship that went from Joppa to Tarshish ; or whether they were alarmed lest a Socinian should undermine that piety which, in the days of Commodore Trunnion, so strikingly characterised sailors, does not appear ; but, at any rate, they objected to Priestley " on account of his religious principles," and appointed the two Forsters, whose " religious principles," if they had been known to these well-meaning but not far-sighted persons, would probably have surprised them.

In 1772 another proposal was made to Priestley.

Lord Shelburne, desiring a "literary companion," had been brought into communication with Priestley by the good offices of a friend of both, Dr. Price; and offered him the nominal post of librarian, with a good house and appointments, and an annuity in case of the termination of the engagement.   Priestley accepted the offer, and remained with Lord Shelburne for seven years, sometimes residing at Calne, sometimes travelling abroad with the Earl.

Why the connection terminated has never been exactly known; but it is certain that Lord Shelburne behaved with the utmost consideration and kindness towards Priestley; that he fulfilled his engagements to the letter; and that, at a later period, he expressed a desire that Priestley should return to his old footing in his house. Probably enough, the politician, aspiring to the highest offices in the State, may have found the position of the protector of a man who was being denounced all over the country as an infidel and an atheist somewhat embarrassing.   In fact, a passage in Priestley's "Autobiography" on the occasion of the publication of his "Disquisitions relating to Matter and Spirit," which took place in 1777, indicates pretty clearly the state of the case :—

"(126) It being probable that this publication would be unpopular, and might be the means of bringing odium on my patron, several attempts were made by his friends, though none by himself, to dissuade me from persisting in it. But being, as I thought, engaged in the cause of important truth, I proceeded without regard to any consequences, assuring them that this publication should not be injurious to his lordship."

It is not unreasonable to suppose that his lordship, as a keen, practical man of the world, did not derive much satisfaction from this assurance. The "evident marks of dissatisfaction" which Priestley says he first perceived in his patron in 1778, may well have arisen from the peer's not unnatural uneasiness as to what his domesticated, but not tamed, philosopher might write next, and what storm might thereby be brought down on his own head; and it speaks very highly for Lord Shelburne's delicacy that, in the midst of such perplexities, he made not the least attempt to interfere with Priestley's freedom of action. In 1780, however, he intimated to Dr. Price that he should be glad to establish Priestley on his Irish estates : the suggestion was interpreted, as Lord Shelburne probably intended it should be, and Priestley left him, the annuity of £150 a year, which had been promised in view of such a contingency, being punctually paid.

After leaving Calne, Priestley spent some little time in London, and then, having settled in Birmingham at the desire of his brother-in-law, he was soon invited to become the minister of a large congregation. This settlement Priestley considered, at the time, to be " the happiest event of

his life." And well he might think so; for it
gave him competence and leisure; placed him
within reach of the best makers of apparatus of
the day; made him a member of that remarkable
" Lunar Society," at whose meetings he could
exchange thoughts with such men as Watt,
Wedgwood, Darwin, and Boulton; and threw
open to him the pleasant house of the Galtons of
Barr, where these men, and others of less note,
formed a society of exceptional charm and intelligence.[1]

But these halcyon days were ended by a bitter
storm. The French Revolution broke out. An
electric shock ran through the nations; whatever
there was of corrupt and retrograde, and, at the
same time, a great deal of what there was of best
and noblest, in European society shuddered at

----

[1] See *The Life of Mary Anne Schimmelpenninck.*" Mrs.
Schimmelpenninck (*née* Galton) remembered Priestley very well,
and her description of him is worth quotation :—" A man of
admirable simplicity, gentleness and kindness of heart, united
with great acuteness of intellect. I can never forget the impression produced on me by the serene expression of his
countenance. He, indeed, seemed present with God by
recollection, and with man by cheerfulness. I remember that,
in the assembly of these distinguished men, amongst whom Mr.
Boulton, by his noble manner, his fine countenance (which much
resembled that of Louis XIV.), and princely munificence, stood
pre-eminently as the great Mecænas ; even as a child, I used to
feel, when Dr. Priestley entered after him, that the glory of the
one was terrestrial, that of the other celestial ; and utterly far
as I am removed from a belief in the sufficiency of Dr.
Priestley's theological creed, I cannot but here record this
evidence of the eternal power of any portion of the truth held
in its vitality."

the outburst of long-pent-up social fires. Men's
feelings were excited in a way that we, in this
generation, can hardly comprehend. Party wrath
and virulence were expressed in a manner un-
paralleled, and it is to be hoped impossible, in our
times; and Priestley and his friends were held up
to public scorn, even in Parliament, as fomenters
of sedition. A " Church-and-King " cry was
raised against the Liberal Dissenters; and, in
Birmingham, it was intensified and specially
directed towards Priestley by a local controversy,
in which he had engaged with his usual vigour.
In 1791, the celebration of the second anniversary
of the taking of the Bastille by a public dinner,
with which Priestley had nothing whatever to do,
gave the signal to the loyal and pious mob, who,
unchecked, and indeed to some extent encouraged,
by those who were responsible for order, had the
town at their mercy for three days. The chapels
and houses of the leading Dissenters were
wrecked, and Priestley and his family had to fly
for their lives, leaving library, apparatus, papers,
and all their possessions, a prey to the flames.

Priestley never returned to Birmingham. He
bore the outrages and losses inflicted upon him
with extreme patience and sweetness,[1] and betook

[1] Even Mrs. Priestley, who might be forgiven for regarding
the destroyers of her household gods with some asperity,
contents herself, in writing to Mrs. Barbauld, with the sarcasm
that the Birmingham people "will scarcely find so many
respectable characters, a second time, to make a bonfire of."

himself to London. But even his scientific colleagues gave him a cold shoulder; and though he
was elected minister of a congregation at Hackney,
he felt his position to be insecure, and finally determined on emigrating to the United States. He
landed in America in 1794; lived quietly with his
sons at Northumberland, in Pennsylvania, where
his posterity still flourish; and, clear-headed and
busy to the last, died on the 6th of February
1804.

Such were the conditions under which Joseph
Priestley did the work which lay before him, and
then, as the Norse Sagas say, went out of the
story. The work itself was of the most varied
kind. No human interest was without its attraction for Priestley, and few men have ever had so
many irons in the fire at once; but, though he
may have burned his fingers a little, very few
who have tried that operation have burned their
fingers so little. He made admirable discoveries
in science; his philosophical treatises are still
well worth reading; his political works are full of
insight and replete with the spirit of freedom; and
while all these sparks flew off from his anvil, the
controversial hammer rained a hail of blows on
orthodox priest and bishop. While thus engaged,
the kindly, cheerful doctor felt no more wrath or
uncharitableness towards his opponents than a
smith does towards his iron. But if the iron

could only speak !—and the priests and bishops took the point of view of the iron.

No doubt what Priestley's friends repeatedly urged upon him—that he would have escaped the heavier trials of his life and done more for the advancement of knowledge, if he had confined himself to his scientific pursuits and let his fellow-men go their way—was true. But it seems to have been Priestley's feeling that he was a man and a citizen before he was a philosopher, and that the duties of the two former positions are at least as imperative as those of the latter. More-over, there are men (and I think Priestley was one of them) to whom the satisfaction of throwing down a triumphant fallacy is as great as that which attends the discovery of a new truth ; who feel better satisfied with the government of the world, when they have been helping Providence by knocking an imposture on the head ; and who care even more for freedom of thought than for mere advance of knowledge. These men are the Carnots who organise victory for truth, and they are, at least, as important as the generals who visibly fight her battles in the field.

Priestley's reputation as a man of science rests upon his numerous and important contributions to the chemistry of gaseous bodies ; and to form a just estimate of the value of his work—of the extent to which it advanced the knowledge of fact and the development of sound theoretical views—we must reflect what chemistry was in the first half of the eighteenth century.

The vast science which now passes under that name had no existence. Air, water, and fire were still counted among the elemental bodies; and though Van Helmont, a century before, had distinguished different kinds of air as *gas ventosum* and *gas sylvestre*, and Boyle and Hales had experimentally defined the physical properties of air, and discriminated some of the various kinds of aëriform bodies, no one suspected the existence of the numerous totally distinct gaseous elements which are now known, or dreamed that the air we breathe and the water we drink are compounds of gaseous elements.

But, in 1754, a young Scotch physician, Dr. Black, made the first clearing in this tangled backwood of knowledge. And it gives one a wonderful impression of the juvenility of scientific chemistry to think that Lord Brougham, whom so many of us recollect, attended Black's lectures when he was a student in Edinburgh. Black's researches gave the world the novel and startling conception of a gas that was a permanently elastic fluid like air, but that differed from common air in being much heavier, very poisonous, and in having the properties of an acid, capable of neutralising the strongest alkalies; and it took the world some time to become accustomed to the notion.

A dozen years later, one of the most sagacious and accurate investigators who has adorned this, or any other, country, Henry Cavendish, published a memoir in the "Philosophical Transactions," in which he deals not only with the "fixed air" (now called carbonic acid or carbonic anhydride) of Black, but with "inflammable air," or what we now term hydrogen.

By the rigorous application of weight and measure to all his processes, Cavendish implied the belief subsequently formulated by Lavoisier, that, in chemical processes, matter is neither created nor destroyed, and indicated the path along which all future explorers must travel. Nor did he himself halt until this path led him, in 1784, to the brilliant and fundamental discovery that water is composed of two gases united in fixed and constant proportions.

It is a trying ordeal for any man to be compared with Black and Cavendish, and Priestley cannot be said to stand on their level. Nevertheless his achievements are not only great in themselves, but truly wonderful, if we consider the disadvantages under which he laboured. Without the careful scientific training of Black, without the leisure and appliances secured by the wealth of Cavendish, he scaled the walls of science as so many Englishmen have done before and since his day; and trusting to mother wit to supply the place of training, and to ingenuity to create apparatus out of washing tubs, he discovered more new gases than all his predecessors put together had done. He laid the foundations of gas analysis; he discovered the complementary actions of animal and vegetable life upon the constituents of the atmosphere; and, finally, he crowned his work, this day one hundred years ago, by the discovery of that " pure dephlogisticated air " to which the French chemists subsequently gave the name of oxygen. Its importance, as the constituent of the atmosphere which disappears in the processes of respiration and combustion, and is restored by green plants growing in sunshine, was proved somewhat later. For these brilliant discoveries, the Royal Society elected Priestley a fellow and gave him their medal, while the Academies of Paris and St. Petersburg conferred their membership upon him. Edinburgh had made him an honorary doctor of laws at an early period of his career; but, I need hardly add, that a man of Priestley's opinions received no recognition from the universities of his own country.

That Priestley's contributions to the knowledge of chemical fact were of the greatest importance, and that they richly deserve all the praise that has been awarded to them, is unquestionable; but it must, at the same time, be admitted that he had no comprehension of the deeper significance of his work; and, so far from contributing anything to the theory of the facts which he discovered, or

assisting in their rational explanation, his influence
to the end of his life was warmly exerted in favour
of error. From first to last, he was a stiff adherent
of the phlogiston doctrine which was prevalent
when his studies commenced ; and, by a curious
irony of fate, the man who by the discovery of
what he called "dephlogisticated air" furnished
the essential datum for the true theory of combustion, of respiration, and of the composition of
water, to the end of his days fought against the
inevitable corollaries from his own labours. His
last. scientific work, published in 1800, bears the
title, "The Doctrine of Phlogiston established, and
that of the Composition of Water refuted."

When Priestley commenced his studies, the current belief was, that atmospheric air, freed from
accidental impurities, is a simple elementary substance, indestructible and unalterable, as water was
supposed to be. When a combustible burned, or
when an animal breathed in air, it was supposed
that a substance, "phlogiston," the matter of heat
and light, passed from the burning or breathing
body into it, and destroyed its powers of supporting
life and combustion. Thus, air contained in a
vessel in which a lighted candle had gone out, or a
living animal had breathed until it could breathe
no longer, was called "phlogisticated." The same
result was supposed to be brought about by the
addition of what Priestley called "nitrous gas" to
common air.

In the course of his researches, Priestley found that the quantity of common air which can thus become "phlogisticated," amounts to about one-fifth the volume of the whole quantity submitted to experiment. Hence it appeared that common air consists, to the extent of four-fifths of its volume, of air which is already "phlogisticated"; while the other fifth is free from phlogiston, or "dephlogisticated." On the other hand, Priestley found that air "phlogisticated" by combustion or respiration could be "dephlogisticated," or have the properties of pure common air restored to it, by the action of green plants in sunshine. The question, therefore, would naturally arise—as common air can be wholly phlogisticated by combustion, and converted into a substance which will no longer support combustion, is it possible to get air that shall be less phlogisticated than common air, and consequently support combustion better than common air does?

Now, Priestley says that, in 1774, the possibility of obtaining air less phlogisticated than common air had not occurred to him.[1] But in pursuing his experiments on the evolution of air from various bodies by means of heat, it happened that, on the 1st of August 1774, he threw the heat of the sun, by means of a large burning glass which he had recently obtained, upon a substance

[1] *Experiments and Observations on Different Kinds of Air*, vol. ii. p. 31.

which was then called *mercurius calcinatus per se,*
and which is commonly known as red precipitate.

"I presently found that, by means of this lens, air was
expelled from it very readily. Having got about three or four
times as much as the bulk of my materials, I admitted wáter
to it, and found that it was not imbibed by it. But what
surprised me more than I can well express, was that a candle
burned in this air with a remarkably vigorous flame, very much
like that enlarged flame with which a candle burns in nitrous
air, exposed to iron or lime of sulphur ; but as I had got
nothing like this remarkable appearance from any kind of air
besides this particular modification of nitrous air, and I knew
no nitrous acid was used in the preparation of *mercurius calcinatus,* I was utterly at a loss how to account for it.

"In this case also, though I did not give sufficient attention
to the circumstance at that time, the flame of the candle,
besides being larger, burned with more splendour and heat than
in that species of nitrous air ; and a piece of red-hot wood
sparkled in it, exactly like paper dipped in a solution of nitre,
and it consumed very fast—an experiment which I had never
thought of trying with nitrous air." [1]

Priestley obtained the same sort of air from red
lead, but, as he says himself, he remained in
ignorance of the properties of this new kind of air
for seven months, or until March 1775, when he
found that the new air behaved with "nitrous
gas" in the same way as the dephlogisticated part
of common air does ; [2] but that, instead of being
diminished to four-fifths, it almost completely
vanished, and, therefore, showed itself to be "between five and six times as good as the best

[1] *Experiments and Observations on Different Kinds of Air,*
vol. ii. pp. 34, 35.
[2] *Ibid.* vol. i. p. 40.

common air I have ever met with." [1] As this new air thus appeared to be completely free from phlogiston, Priestley called it "dephlogisticated air."

What was the nature of this air? Priestley found that the same kind of air was to be obtained by moistening with the spirit of nitre (which he terms nitrous acid) any kind of earth that is free from phlogiston, and applying heat; and consequently he says: "There remained no doubt on my mind but that the atmospherical air, or the thing that we breathe, consists of the nitrous acid and earth, with so much phlogiston as is necessary to its elasticity, and likewise so much more as is required to bring it from its state of perfect purity to the mean condition in which we find it." [2]

Priestley's view, in fact, is that atmospheric air is a kind of saltpetre, in which the potash is replaced by some unknown earth. And in speculating on the manner in which saltpetre is formed, he enunciates the hypothesis, "that nitre is formed by a real *decomposition of the air itself*, the *bases* that are presented to it having, in such circumstances, a nearer affinity with the spirit of nitre than that kind of earth with which it is united in the atmosphere." [3]

---

[1] *Experiments and Observations on Different Kinds of Air*, vol. ii. p. 48.   [2] *Ibid.* p. 55.
[3] *Ibid.* p. 60. The italics are Priestley's own.

It would have been hard for the most ingenious person to have wandered farther from the truth than Priestley does in this hypothesis ; and, though Lavoisier undoubtedly treated Priestley very ill, and pretended to have discovered dephlogisticated air, or oxygen, as he called it, independently, we can almost forgive him when we reflect how different were the ideas which the great French chemist attached to the body which Priestley discovered.

They are like two navigators of whom the first sees a new country, but takes clouds for mountains and mirage for lowlands ; while the second determines its length and breadth, and lays down on a chart its exact place, so that, thenceforth, it serves as a guide to his successors, and becomes a secure outpost whence new explorations may be pushed.

Nevertheless, as Priestley himself somewhere remarks, the first object of physical science is to ascertain facts, and the service which he rendered to chemistry by the definite establishment of a large number of new and fundamentally important facts, is such as to entitle him to a very high place among the fathers of chemical science.

It is difficult to say whether Priestley's philosophical, political, or theological views were most responsible for the bitter hatred which was borne to him by a large body of his country-

men,[1] and which found its expression in the
malignant insinuations in which Burke, to his
everlasting shame, indulged in the House of
Commons.

Without containing much that will be new to
the readers of Hobbs, Spinoza, Collins, Hume, and
Hartley, and, indeed, while making no pretensions
to originality, Priestley's " Disquisitions relating
to Matter and Spirit," and his " Doctrine of Philosophical Necessity Illustrated," are among the
most powerful, clear, and unflinching expositions
of materialism and necessarianism which exist in
the English language, and are still well worth
reading.

Priestley denied the freedom of the will in the
sense of its self-determination; he denied the
existence of a soul distinct from the body ; and as
a natural consequence, he denied the natural immortality of man.

In relation to these matters English opinion, a
century ago, was very much what it is now.

---

[1] " In all the newspapers and most of the periodical publications I was represented as an unbeliever in Revelation, and no
better than an atheist."—*Autobiography*, Rutt, vol i. p. 124.
"On the walls of houses, etc., and especially where I usually
went, were to be seen, in large characters, ' MADAN FOR EVER ;
DAMN PRIESTLEY ; NO PRESBYTERIANISM ; DAMN THE PRES-
BYTERIANS,' etc., etc. ; and, at one time, I was followed by a
number of boys, who left their play, repeating what they had
seen on the walls, and shouting out, ' *Damn Priestley ; damn
him, damn him, for ever, for ever,*' etc., etc. This was no
doubt a lesson which they had been taught by their parents,
and what they, I fear, had learned from their superiors."—
Appeal to the Public on the Subject of the Riots at Birmingham.

A man may be a necessarian without incurring graver reproach than that implied in being called a gloomy fanatic, necessarianism, though very shocking, having a note of Calvanistic orthodoxy; but, if a man is a materialist; or, if good authorities say he is and must be so, in spite of his assertion to the contrary; or, if he acknowledge himself unable to see good reasons for believing in the natural immortality of man, respectable folks look upon him as an unsafe neighbour of a cash-box, as an actual or potential sensualist, the more virtuous in outward seeming, the more certainly loaded with secret " grave personal sins."

Nevertheless, it is as certain as anything can be, that Joseph Priestley was no gloomy fanatic, but as cheerful and kindly a soul as ever breathed, the idol of children; a man who was hated only by those who did not know him, and who charmed away the bitterest prejudices in personal inter-course; a man who never lost a friend, and the best testimony to whose worth is the generous and tender warmth with which his many friends vied with one another in rendering him substantial help, in all the crises of his career.

The unspotted purity of Priestley's life, the strictness of his performance of every duty, his transparent sincerity, the unostentatious and deep-seated piety which breathes through all his correspondence, are in themselves a sufficient refutation of the hypothesis, invented by bigots to cover uncharitableness, that such opinions as his must arise from moral defects. And his statue will do as good service as the brazen image that was set upon a pole before the Israelites, if those who have been bitten by the fiery serpents of sectarian hatred, which still haunt this wilderness of a world, are made whole by looking upon the image of a heretic who was yet a saint.

Though Priestley did not believe in the natural immortality of man, he held with an almost naïve realism that man would be raised from the dead by a direct exertion of the power of God, and thenceforward be immortal. And it may be as well for those who may be shocked by this doctrine to know that views, substantially identical with Priestley's, have been advocated, since his time, by two prelates of the Anglican Church : by Dr. Whately, Archbishop of Dublin, in his well-known " Essays "; [1] and by Dr. Courtenay, Bishop of Kingston in Jamaica, the first edition of whose remarkable book " On the Future States," dedicated to Archbishop Whately, was published in 1843 and the second in 1857. According to Bishop Courtenay,

" The death of the body will cause a cessation of all the activity of the mind by way of natural consequence ; to continue for ever UNLESS the Creator should interfere."

----

[1] First Series. *On Some of the Peculiarities of the Christian Religion.* Essay I. "Revelation of a Future State."

And again :—

"The natural end of human existence is the 'first death, the dreamless slumber of the grave, wherein man lies spell-bound, soul and body, under the dominion of sin and death—that whatever modes of conscious existence, whatever future states of 'life' or of 'torment' beyond Hades are reserved for man, are results of our blessed Lord's victory over sin and death ; that the resurrection of the dead must be preliminary to their entrance into either of the future states, and that the nature and even existence of these states, and even the mere fact that there is a futurity of consciousness, can be known *only* through God's revelation of Himself in the Person and the Gospel of His Son."—P. 389.

And now hear Priestley :—

"Man, according to this system (of materialism), is no more than we now see of him. His being commences at the time of his conception, or perhaps at an earlier period. The corporeal and mental faculties, in being in the same substance, grow, ripen, and decay together; and whenever the system is dissolved it continues in a state of dissolution till it shall please that Almighty Being who called it into existence to restore it to life again."—"Matter and Spirit," p. 49.

And again :—

"The doctrine of the Scripture is, that God made man of the dust of the ground, and by simply animating this organised matter, made man that living percipient and intelligent being that he is. According to Revelation, *death* is a state of rest and insensibility, and our only though sure hope of a future life is founded on the doctrine of the resurrection of the whole man at some distant period ; this assurance being sufficiently confirmed to us both by the evident tokens of a Divine commission attending the persons who delivered the doctrine, and especially by the actual resurrection of Jesus Christ, which is more authentically attested than any other fact in history."—*Ibid.*, p. 247.

We all know that " a saint in crape is twice a saint in lawn ; " but it is not yet admitted that the views which are consistent with such saintliness in lawn, become diabolical when held by a mere dissenter.[1]

I am not here either to defend or to attack Priestley's philosophical views, and I cannot say that I am personally disposed to attach much value to episcopal authority in philosophical questions; but it seems right to call attention to the fact, that those of Priestley's opinions which have brought most odium upon him have been openly promulgated, without challenge, by persons occupying the highest positions in the State Church.

I must confess that what interests me most about Priestley's materialism, is the evidence that he saw dimly the seed of destruction which such materialism carries within its own bosom. In the course of his reading for his " History of Discoveries relating to Vision, Light, and Colours," he had come upon the speculations of Boscovich

---

[1] Not only is Priestley at one with Bishop Courtenay in this matter, but with Hartley and Bonnet, both of them stout champions of Christianity. Moreover, Archbishop Whately's essay is little better than an expansion of the first paragraph of Hume's famous essay on the Immortality of the Soul :—" By the mere light of reason it seems difficult to prove the immortality of the soul ; the arguments for it are commonly derived either from metaphysical topics, or moral, or physical. But it is in reality the Gospel, and the Gospel alone, that has brought *life and immortality to light.*" It is impossible to imagine that a man of Whately's tastes and acquirements had not read Hume or Hartley, though he refers to neither.

and Michell, and had been led to admit the sufficiently obvious truth that our knowledge of matter
is a knowledge of its properties; and that of its
substance—if it have a substance—we know nothing. And this led to the further admission that,
so far as we can know, there may be no difference
between the substance of matter and the substance
of spirit (" Disquisitions," p. 16). A step farther
would have shown Priestley that his materialism
was, essentially, very little different from the
Idealism of his contemporary, the Bishop of Cloyne.

As Priestley's philosophy is mainly a clear statement of the views of the deeper thinkers of his day,
so are his political conceptions based upon those of
Locke. Locke's aphorism that " the end of government is the good of mankind," is thus expanded by
Priestley :—

" It must necessarily be understood, therefore, whether it be
expressed or not, that all people live in society for their mutual
advantage ; so that the good and happiness of the members,
that is, of the majority of the members, of any state, is the
great standard by which everything relating to that state must
finally be determined." [1]

The little sentence here interpolated, " that is,
of the majority of the members of any state," appears to be that passage which suggested to
Bentham, according to his own acknowledgment,
the famous " greatest happiness " formula, which

[1] *Essay on the First Principles of Government.* Second
edition, 1771, p. 13.

by substituting " happiness " for " good," has converted a noble into an ignoble principle.   But I do
not call to mind that there is any utterance in
Locke quite so outspoken as the following passage
in the "Essay on the First Principles of Government."   After laying down as "a fundamental
maxim in all Governments," the proposition that
" kings, senators, and nobles " are " the servants
of the public," Priestley goes on to say :—

> " But in the largest states, if the abuses of the government
> should at any time be great and manifest ; if the servants of
> the people, forgetting their masters and their masters' interest,
> should pursue a separate one of their own ; if, instead of con-
> sidering that they are made for the people, they should consider
> the people as made for them ; if the oppressions and violation
> of right should be great, flagrant, and universally resented ; if
> the tyrannical governors should have no friends but a few
> sycophants, who had long preyed upon the vitals of their fellow-
> citizens, and who might be expected to desert a government
> whenever their interests should be detached from it : if, in
> consequence of these circumstances, it should become manifest
> that the risk which would be run in attempting a revolution
> would be trifling, and the evils which might be apprehended
> from it were far less than those which were actually suffered
> and which were daily increasing ; in the name of God, I ask,
> what principles are those which ought to restrain an injured and
> insulted people from asserting their natural rights, and from
> changing or even punishing their governors—that is, their
> servants—who had abused their trust, or from altering the
> whole form of their government, if it appeared to be of a struc-
> ture so liable to abuse ?"

As a Dissenter, subject to the operation of the
Corporation and Test Acts, and as a Unitarian
excluded from the benefit of the Toleration Act,

it is not surprising to find that Priestley had very definite opinions about Ecclesiastical Establishments ; the only wonder is that these opinions were so moderate as the following passages show them to have been :—

"Ecclesiastical authority may have been necessary in the infant state of society, and, for the same reason, it may perhaps continue to be, in some degree, necessary as long as society is imperfect ; and therefore may not be entirely abolished till civil governments have arrived at a much greater degree of perfection. If, therefore, I were asked whether I should approve of the immediate dissolution of all the ecclesiastical establishments in Europe, I should answer, No. . . . Let experiment be first made of *alterations*, or, which is the same thing, of *better establishments* than the present. Let them be reformed in many essential articles, and then not thrown aside entirely till it be found by experience that no good can be made of them."

Priestley goes on to suggest four such reforms of a capital nature :—

"1. Let the Articles of Faith to be subscribed by candidates for the ministry be greatly reduced. In the formulary of the Church of England, might not thirty-eight out of the thirty-nine be very well spared ? It is a reproach to any Christian establishment if every man cannot claim the benefit of it who can say that he believes in the religion of Jesus Christ as it is set forth in the New Testament. You say the terms are so general that even Deists would quibble and insinuate themselves. I answer that all the articles which are subscribed at present by no means exclude Deists who will prevaricate ; and upon this scheme you would at least exclude fewer honest men." [1]

---

[1] "Utility of Establishments," in *Essay on First Principles of Government*, p. 198, 1771.

The second reform suggested is the equalisa-
tion, in proportion to work done, of the stipends
of the clergy; the third, the exclusion of the
Bishops from Parliament; and the fourth, complete toleration, so that every man may enjoy the
rights of a citizen, and be qualified to serve his
country, whether he belong to the Established
Church or not.

Opinions such as those I have quoted, respecting
the duties and the responsibilities of governors,
are the commonplaces of modern Liberalism;
and Priestley's views on Ecclesiastical Establishments would, I fear, meet with but a cool reception, as altogether too conservative, from a
large proportion of the lineal descendants of the
people who taught their children to cry "Damn
Priestley;" and with that love for the practical
application of science which is the source of the
greatness of Birmingham, tried to set fire to the
doctor's house with sparks from his own electrical
machine; thereby giving the man they called an
incendiary and raiser of sedition against Church
and King, an appropriately experimental illustration of the nature of arson and riot.

If I have succeeded in putting before you the
main features of Priestley's work, its value will
become apparent when we compare the condition
of the English nation, as he knew it, with its
present state.

The fact that France has been for eighty-five years trying, without much success, to right herself after the great storm of the Revolution, is not unfrequently cited among us as an indication of some inherent incapacity for self-government among the French people. I think, however, that Englishmen who argue thus, forget that, from the meeting of the Long Parliament in 1640, to the last Stuart rebellion in 1745, is a hundred and five years, and that, in the middle of the last century, we had but just safely freed ourselves from our Bourbons and all that they represented. The corruption of our state was as bad as that of the Second Empire. Bribery was the instrument of government, and peculation its reward. Four-fifths of the seats in the House of Commons were more or less openly dealt with as property. A minister had to consider the state of the vote market, and the sovereign secured a sufficiency of "king's friends" by payments allotted with retail, rather than royal, sagacity.

Barefaced and brutal immorality and intemperance pervaded the land, from the highest to the lowest classes of society. The Established Church was torpid, as far as it was not a scandal; but those who dissented from it came within the meshes of the Act of Uniformity, the Test Act, and the Corporation Act. By law, such a man as Priestley, being a Unitarian, could neither teach nor preach, and was liable to ruinous fines and

long imprisonment.[1] In those days the guns
that were pointed by the Church against the
Dissenters were shotted. The law was a cesspool
of iniquity and cruelty. Adam Smith was a new
prophet whom few regarded, and commerce was
hampered by idiotic impediments, and ruined
by still more absurd help, on the part of
government.

Birmingham, though already the centre of a
considerable industry, was a mere village as
compared with its present extent. People who
travelled went about armed, by reason of the
abundance of highwaymen and the paucity and
inefficiency of the police. Stage coaches had
not reached Birmingham, and it took three days
to get to London. Even canals were a recent
and much opposed invention.

Newton had laid the foundation of a mechanical
conception of the physical universe : Hartley,
putting a modern face upon ancient materialism,
had extended that mechanical conception to psychology; Linnæus and Haller were beginning to
introduce method and order into the chaotic
acccumulation of biological facts. But those
parts of physical science which deal with heat,
electricity, and magnetism, and above all,
chemistry, in the modern sense, can hardly
be said to have had an existence. No one

---

[1] In 1732 Doddridge was cited for teaching without the
Bishop's leave, at Northampton.

knew that two of the old elemental bodies, air and water, are compounds, and that a third, fire, is not a substance but a motion. The great industries that have grown out of the applications of modern scientific discoveries had no existence, and the man who should have foretold their coming into being in the days of his son, would have been regarded as a mad enthusiast.

In common with many other excellent persons, Priestley believed that man is capable of reaching, and will eventually attain, perfection. If the temperature of space presented no obstacle, I should be glad to entertain the same idea; but judging from the past progress of our species, I am afraid that the globe will have cooled down so far, before the advent of this natural millennium, that we shall be, at best, perfected Esquimaux. For all practical purposes, however, it is enough that man may visibly improve his condition in the course of a century or so. And, if the picture of the state of things in Priestley's time, which I have just drawn, have any pretence to accuracy, I think it must be admitted that there has been a considerable change for the better.

I need not advert to the well-worn topic of material advancement, in a place in which the very stones testify to that progress—in the town of Watt and of Boulton. I will only remark, in passing, that material advancement has its share in moral and intellectual progress. Becky Sharp's acute remark that it is not difficult to be virtuous on ten thousand a year, has its application to nations; and it is futile to expect a hungry and squalid population to be anything but violent and gross. But as regards other than material welfare, although perfection is not yet in sight— even from the mast-head—it is surely true that things are much better than they were.

Take the upper and middle classes as a whole, and it may be said that open immorality and gross intemperance have vanished. Four and six bottle men are as extinct as the dodo. Women of good repute do not gamble, and talk modelled upon Dean Swift's "Art of Polite Conversation" would be tolerated in no decent kitchen.

Members of the legislature are not to be bought; and constituents are awakening to the fact that votes must not be sold—even for such trifles as rabbits and tea and cake. Political power has passed into the hands of the masses of the people. Those whom Priestley calls their servants have recognised their position, and have requested the master to be so good as to go to school and fit himself for the administration of his property. In ordinary life, no civil disability attaches to any one on theological grounds, and high offices of the state are open to Papist, Jew, and Secularist.

Whatever men's opinions as to the policy of Establishment, no one can hesitate to admit that

the clergy of the Church are men of pure life
and conversation, zealous in the discharge of their
duties ; and at present, apparently, more bent on
prosecuting one another than on meddling with
Dissenters. Theology itself has broadened so
much, that Anglican divines put forward doctrines
more liberal than those of Priestley; and, in our
state-supported churches, one listener may hear a
sermon to which Bossuet might have given his
approbation, while another may hear a discourse
in which Socrates would find nothing new.

But great as these changes may be, they sink
into insignificance beside the progress of physical
science, whether we consider the improvement of
methods of investigation, or the increase in bulk
of solid knowledge. Consider that the labours of
Laplace, of Young, of Davy, and of Faraday; of
Cuvier, of Lamarck, and of Robert Brown; of
Von Baer, and of Schwann; of Smith and of
Hutton, have all been carried on since Priestley
discovered oxygen ; and consider that they are
now things of the past, concealed by the industry
of those who have built upon them, as the first
founders of a coral reef are hidden beneath the
life's work of their successors ; consider that the
methods of physical science are slowly spreading
into all investigations, and that proofs as valid as
those required by her canons of investigation are
being demanded of all doctrines which ask for
men's assent; and you will have a faint image of

the astounding difference in this respect between the nineteenth century and the eighteenth.

If we ask what is the deeper meaning of all these vast changes, I think there can be but one reply. They mean that reason has asserted and exercised her primacy over all provinces of human activity: that ecclesiastical authority has been relegated to its proper place; that the good of the governed has been finally recognised as the end of government, and the complete responsibility of governors to the people as its means; and that the dependence of natural phenomena in general on the laws of action of what we call matter has become an axiom.

But it was to bring these things about, and to enforce the recognition of these truths, that Joseph Priestley laboured. If the nineteenth century is other and better than the eighteenth, it is, in great measure, to him, and to such men as he, that we owe the change. If the twentieth century is to be better than the nineteenth, it will be because there are among us men who walk in Priestley's footsteps.

Such men are not those whom their own generation delights to honour; such men, in fact, rarely trouble themselves about honour, but ask, in another spirit than Falstaff's, "What is honour? Who hath it? He that died o' Wednesday." But whether Priestley's lot be theirs, and a future generation, in justice and in gratitude, set up

their statues; or whether their names and fame
are blotted out from remembrance, their work will
live as long as time endures.   To all eternity, the
sum of truth and right will have been increased
by their means; to all eternity, falsehood and
injustice will be the weaker because they have
lived.

II

ON THE EDUCATIONAL VALUE OF THE NATURAL HISTORY SCIENCES

[1854.]

THE subject to which I have to beg your attention during the ensuing hour is " The Relation of Physiological Science to other branches of Knowledge."

Had circumstances permitted of the delivery, in their strict logical order, of that series of discourses of which the present lecture is a member, I should have preceded my friend and colleague Mr. Henfrey, who addressed you on Monday last ; but while, for the sake of that order, I must beg you to suppose that this discussion of the Educational bearings of Biology in general *does* precede that of Special Zoology and Botany, I am rejoiced to be able to take advantage of the light thus already thrown upon the tendency and methods of Physiological Science.

Regarding Physiological Science, then, in its

widest sense—as the equivalent of *Biology*—the Science of Individual Life—we have to consider in succession :

1. Its position and scope as a branch of knowledge.

2. Its value as a means of mental discipline.

3. Its worth as practical information.

And lastly,

4. At what period it may best be made a branch of Education.

Our conclusions on the first of these heads must depend, of course, upon the nature of the subject-matter of Biology ; and I think a few preliminary considerations will place before you in a clear light the vast difference which exists between the living bodies with which Physiological science is concerned, and the remainder of the universe;—between the phænomena of Number and Space, of Physical and of Chemical force, on the one hand, and those of Life on the other.

The mathematician, the physicist, and the chemist contemplate things in a condition of rest; they look upon a state of equilibrium as that to which all bodies normally tend.

The mathematician does not suppose that a quantity will alter, or that a given point in space will change its direction with regard to another point, spontaneously. And it is the same with the physicist. When Newton saw the apple fall, he concluded at once that the act of falling was not the result of any power inherent in the apple, but that it was the result of the action of something else on the apple. In a similar manner, all physical force is regarded as the disturbance of an equilibrium to which things tended before its exertion,—to which they will tend again after its cessation.

The chemist equally regards chemical change in a body as the effect of the action of something external to the body changed. A chemical compound once formed would persist for ever, if no alteration took place in surrounding conditions.

But to the student of Life the aspect of Nature is reversed. Here, incessant, and, so far as we know, spontaneous change is the rule, rest the exception—the anomaly to be accounted for. Living things have no inertia, and tend to no equilibrium.

Permit me, however, to give more force and clearness to these somewhat abstract considerations by an illustration or two.

Imagine a vessel full of water, at the ordinary temperature, in an atmosphere saturated with vapour. The *quantity* and the *figure* of that water will not change, so far as we know, for ever.

Suppose a lump of gold be thrown into the vessel —motion and disturbance of figure exactly proportional to the momentum of the gold will take place. But after a time the effects of this disturbance will subside—equilibrium will be restored, and the water will return to its passive state.

Expose the water to cold—it will solidify—and in so doing its particles will arrange themselves in definite crystalline shapes. But once formed, these crystals change no further.

Again, substitute for the lump of gold some substance capable of entering into chemical relations with the water:—say, a mass of that substance which is called "protein"—the substance of flesh:—a very considerable disturbance of equilibrium will take place—all sorts of chemical compositions and decompositions will occur; but in the end, as before, the result will be the resumption of a condition of rest.

Instead of such a mass of *dead* protein, however, take a particle of *living* protein—one of those minute microscopic living things which throng our pools, and are known as Infusoria—such a creature, for instance, as an Euglena, and place it in our vessel of water. It is a round mass provided with a long filament, and except in this peculiarity of shape, presents no appreciable physical or chemical difference whereby it might be distinguished from the particle of dead protein.

But the difference in the phænomena to which it will give rise is immense : in the first place it will develop a vast quantity of physical force— cleaving the water in all directions with considerable rapidity by means of the vibrations of the long filament or cilium.

Nor is the amount of chemical energy which the little creature possesses less striking. It is a perfect laboratory in itself, and it will act and re-act upon the water and the matters contained therein; converting them into new compounds resembling its own substance, and at the same time giving up portions of its own substance which have become effete.

Furthermore, the Euglena will increase in size; but this increase is by no means unlimited, as the increase of a crystal might be. After it has grown to a certain extent it divides, and each portion assumes the form of the original, and proceeds to repeat the process of growth and division.

Nor is this all. For after a series of such divisions and subdivisions, these minute points assume a totally new form, lose their long tails—round themselves, and secrete a sort of envelope or box, in which they remain shut up for a time, eventually to resume, directly or indirectly, their primitive mode of existence.

Now, so far as we know, there is no natural limit to the existence of the Euglena, or of any other living germ. A living species once launched into existence tends to live for ever.

Consider how widely different this living particle is from the dead atoms with which the physicist and chemist have to do!

The particle of gold falls to the bottom and rests—the particle of dead protein decomposes and disappears—it also rests: but the *living* protein mass neither tends to exhaustion of its forces nor to any permanency of form, but is essentially distinguished as a disturber of equilibrium so far as force is concerned,—as undergoing continual metamorphosis and change, in point of form.

Tendency to equilibrium of force and to permanency of form, then, are the characters of that portion of the universe which does not live— the domain of the chemist and physicist.

Tendency to disturb existing equilibrium—to take on forms which succeed one another in definite cycles—is the character of the living world.

What is the cause of this wonderful difference between the dead particle and the living particle of matter appearing in other respects identical? that difference to which we give the name of Life?

I, for one, cannot tell you. It may be that, by and by, philosophers will discover some higher laws of which the facts of life are particular cases —very possibly they will find out some bond between physico-chemical phænomena on the one hand, and vital phænomena on the other. At present, however, we assuredly know of none; and I think we shall exercise a wise humility in confessing that, for us at least, this successive assumption of different states—(external conditions

remaining the same)—this *spontaneity of action*—
if I may use a term which implies more than I
would be answerable for—which constitutes so
vast and plain a practical distinction between
living bodies and those which do not live, is an ultimate fact; indicating as such, the existence of a
broad line of demarcation between the subject-
matter of Biological and that of all other
sciences.

For I would have it understood that this simple
Euglena is the type of *all* living things, so far as
the distinction between these and inert matter is
concerned.   That cycle of changes, which is constituted by perhaps not more than two or three
steps in the Euglena, is as clearly manifested in
the multitudinous stages through which the germ
of an oak or of a man passes.   Whatever forms
the Living Being may take on, whether simple or
complex, *production, growth, reproduction,* are the
phænomena which distinguish it from that which
does not live.

If this be true, it is clear that the student, in
passing from the physico-chemical to the physiological sciences, enters upon a totally new order of
facts; and it will next be for us to consider how
far these new facts involve *new* methods, or require
a modification of those with which he is already
acquainted.   Now a great deal is said about the
peculiarity of the scientific method in general, and
of the different methods which are pursued in the

different sciences.   The Mathematics are said to
have one special method ; Physics another, Biology
a third, and so forth.   For my own part, I must
confess that I do not understand this phraseology.

So far as I can arrive at any clear comprehension
of the matter, Science is not, as many would seem
to suppose, a modification of the black art, suited
to the tastes of the nineteenth century, and flour-
ishing mainly in consequence of the decay of the
Inquisition.

Science is, I believe, nothing but *trained and
organised common sense*, differing from the latter
only as a veteran may differ from a raw recruit: and
its methods differ from those of common sense
only so far as the guardsman's cut and thrust differ
from the manner in which a savage wields his club.
The primary power is the same in each case, and
perhaps the untutored savage has the more brawny
arm of the two.   The *real* advantage lies in the
point and polish of the swordsman's weapon ; in
the trained eye quick to spy out the weakness of
the adversary; in the ready hand prompt to follow
it on the instant.   But, after all, the sword exercise is only the hewing and poking of the clubman
developed and perfected.

So, the vast results obtained by Science are won
by no mystical faculties, by no mental processes,
other than those which are practised by every one
of us, in the humblest and meanest affairs of life.
A detective policeman discovers a burglar from the

marks made by his shoe, by a mental process identical with that by which Cuvier restored the extinct animals of Montmartre from fragments of their bones.   Nor does that process of induction and deduction by which a lady, finding a stain of a peculiar kind upon her dress, concludes that somebody has upset the inkstand thereon, differ in any way, in kind, from that by which Adams and Leverrier discovered a new planet.

The man of science, in fact, simply uses with scrupulous exactness the methods which we all, habitually and at every moment, use carelessly ; and the man of business must as much avail himself of the scientific method—must be as truly a man of science—as the veriest bookworm of us all ; though I have no doubt that the man of business will find himself out to be a philosopher with as much surprise as M. Jourdain exhibited when he discovered that he had been all his life talking prose.   If, however, there be no real difference between the methods of science and those of common life, it would seem, on the face of the matter, highly improbable that there should be any difference between the methods of the different sciences ; nevertheless, it is constantly taken for granted that there is a very wide difference between the Physiological and other sciences in point of method.

In the first place it is said—and I take this point first, because the imputation is too frequently admitted by Physiologists themselves—that Biology differs from the Physico-chemical and Mathematical sciences in being "inexact."

Now, this phrase "inexact" must refer either to the *methods* or to the *results* of Physiological science.

It cannot be correct to apply it to the methods; for, as I hope to show you by and by, these are identical in all sciences, and whatever is true of Physiological method is true of Physical and Mathematical method.

Is it then the *results* of Biological science which are "inexact"? I think not. If I say that respiration is performed by the lungs; that digestion is effected in the stomach; that the eye is the organ of sight; that the jaws of a vertebrated animal never open sideways, but always up and down; while those of an annulose animal always open sideways, and never up and down—I am enumerating propositions which are as exact as anything in Euclid. How then has this notion of the inexactness of Biological science come about? I believe from two causes: first, because in consequence of the great complexity of the science and the multitude of interfering conditions, we are very often only enabled to predict approximately what will occur under given circumstances; and secondly, because, on account of the comparative youth of the Physiological sciences, a great many of their laws are still imperfectly worked out.

But, in an educational point of view, it is most important to distinguish between the essence of a science and the accidents which surround it; and essentially, the methods and results of Physiology are as exact as those of Physics or Mathematics.

It is said that the Physiological method is especially *comparative* [1]; and this dictum also finds favour in the eyes of many. I should be sorry to suggest that the speculators on scientific classification have been misled by the accident of the name of one leading branch of Biology—*Comparative Anatomy;* but I would ask whether *comparison,* and that classification which is the result of comparison, are not the essence of every science whatsoever? How is it possible to discover a relation of cause and effect of *any* kind without comparing a series of cases together in which the supposed cause and effect occur singly, or combined?

---

[1] "In the third place, we have to review the method of Comparison, which is so specially adapted to the study of living bodies, and by which, above all others, that study must be advanced. In Astronomy, this method is necessarily inapplicable; and it is not till we arrive at Chemistry that this third means of investigation can be used, and then only in subordination to the two others. It is in the study, both statical and dynamical, of living bodies that it first acquires its full development; and its use elsewhere can be only through its application here."—COMTE'S *Positive Philosophy,* translated by Miss Martineau. Vol. i. p. 372.

By what method does M. Comte suppose that the equality or inequality of forces and quantities and the dissimilarity or similarity of forms—points of some slight importance not only in Astronomy and Physics, but even in Mathematics—are ascertained, if not by Comparison?

So far from comparison being in any way peculiar to Biological science, it is, I think, the essence of every science.

A speculative philosopher again tells us that the Biological sciences are distinguished by being sciences of observation and not of experiment![1]

Of all the strange assertions into which speculation without practical acquaintance with a subject may lead even an able man, I think this is the very strangest. Physiology not an experimental science? Why, there is not a function of a single organ in the body which has not been determined wholly and solely by experiment? How did Harvey determine the nature of the circulation, except by experiment? How did Sir Charles Bell determine the functions of the roots of the spinal nerves, save by experiment? How do we know the use of a nerve at all, except by experiment? Nay, how do you know even that your eye is your seeing apparatus, unless you make the experiment of shutting it ; or that your ear is

---

[1] "Proceeding to the second class of means,—Experiment cannot but be less and less decisive, in proportion to the complexity of the phænomena to be explored ; and therefore we saw this resource to be less effectual in chemistry than in physics : and we now find that it is eminently useful in chemistry in comparison with physiology. *In fact, the nature of the phænomena seems to offer almost insurmountable impediments to any extensive and prolific application of such a procedure in biology.*"—COMTE, vol. i. p. 367.

M. Comte, as his manner is, contradicts himself two pages further on, but that will hardly relieve him from the responsibility of such a paragraph as the above.

your hearing apparatus, unless you close it up and thereby discover that you become deaf?

It would really be much more true to say that Physiology is *the* experimental science *par excellence* of all sciences; that in which there is least to be learnt by mere observation, and that which affords the greatest field for the exercise of those faculties which characterise the experimental philosopher. I confess, if any one were to ask me for a model application of the logic of experiment, I should know no better work to put into his hands than Bernard's late Researches on the Functions of the Liver.[1]

Not to give this lecture a too controversial tone, however, I must only advert to one more doctrine, held by a thinker of our own age and country, whose opinions are worthy of all respect. It is, that the Biological sciences differ from all others, inasmuch as in *them* classification takes place by type and not by definition.[2]

It is said, in short, that a natural-history class is not capable of being defined—that the class

[1] *Nouvelle Fonction du Foie considéré comme organe producteur de matière sucrée chez l'Homme et les Animaux,* par M. Claude Bernard.

[2] "*Natural Groups given by Type, not by Definition.* . . . . The class is steadily fixed, though not precisely limited; it is given, though not circumscribed; it is determined, not by a boundary-line without, but by a central point within; not by what it strictly excludes, but what it eminently includes; by an example, not by a precept; in short, instead of Definition we have a *Type* for our director. A type is an example of any class, for instance, a species of a genus, which is considered as

Rosaceæ, for instance, or the class of Fishes, is not accurately and absolutely definable, inasmuch as its members will present exceptions to every possible definition; and that the members of the class are united together only by the circumstance that they are all more like some imaginary average rose or average fish, than they resemble anything else.

But here, as before, I think the distinction has arisen entirely from confusing a transitory imperfection with an essential character. So long as our information concerning them is imperfect, we class all objects together according to resemblances which we *feel*, but cannot *define;* we group them round *types*, in short. Thus if you ask an ordinary person what kinds of animals there are, he will probably say, beasts, birds, reptiles, fishes, insects, &c. Ask him to define a beast from a reptile, and he cannot do it; but he says, things like a cow or a horse are beasts, and things like a frog or a lizard are reptiles. You see *he does* class by type, and not by definition. But how does this classification differ from that of the scientific Zoologist? How does the meaning of the scientific class-name of " Mammalia " differ from the unscientific of " Beasts " ?

eminently possessing the characters of the class. All the species which have a greater affinity with this type-species than with any others, form the genus, and are ranged about it, deviating from it in various directions and different degrees."—WHE-WELL, *The Philosophy of the Inductive Sciences*, vol. i. pp. 476, 477.

Why, exactly because the former depends on a definition, the latter on a type. The class Mammalia is scientifically defined as " all animals which have a vertebrated skeleton and suckle their young." Here is no reference to type, but a definition rigorous enough for a geometrician. And such is the character which every scientific naturalist recognises as that to which his classes must aspire—knowing, as he does, that classification by type is simply an acknowledgment of ignorance and a temporary device.

So much in the way of negative argument as against the reputed differences between Biological and other methods. No such differences, I believe, really exist. The subject-matter of Biological science is different from that of other sciences, but the methods of all are identical; and these methods are—

1. *Observation* of facts—including under this head that *artificial observation* which is called *experiment*.

2. That process of tying up similar facts into bundles, ticketed and ready for use, which is called *Comparison* and *Classification*,—the results of the process, the ticketed bundles, being named *General propositions*.

3. *Deduction*, which takes us from the general proposition to facts again—teaches us, if I may so say, to anticipate from the ticket what is inside the bundle. And finally—

4. *Verification,* which is the process of ascertaining whether, in point of fact, our anticipation is a correct one.

Such are the methods of all science whatsoever; but perhaps you will permit me to give you an illustration of their employment in the science of Life; and I will take as a special case the establishment of the doctrine of the *Circulation of the Blood.*

In this case, *simple observation* yields us a knowledge of the existence of the blood from some accidental hæmorrhage, we will say ; we may even grant that it informs us of the localisation of this blood in particular vessels, the heart, &c., from some accidental cut or the like.   It teaches also the existence of a pulse in various parts of the body, and acquaints us with the structure of the heart and vessels.

Here, however, *simple observation* stops, and we must have recourse to *experiment.*

You tie a vein, and you find that the blood accumulates on the side of the ligature opposite the heart.   You tie an artery, and you find that the blood accumulates on the side near the heart. Open the chest, and you see the heart contracting with great force.   Make openings into its principal cavities, and you will find that all the blood flows out, and no more pressure is exerted on either side of the arterial or venous ligature.

Now all these facts, taken together, constitute

the evidence that the blood is propelled by the heart through the arteries, and returns by the veins —that, in short, the blood circulates.

Suppose our experiments and observations have been made on horses, then we group and ticket them into a general proposition, thus :—*all horses have a circulation of their blood.*

Henceforward a horse is a sort of indication or label, telling us where we shall find a peculiar series of phænomena called the circulation of the blood.

Here is our *general proposition*, then.

How, and when, are we justified in making our next step—a *deduction* from it ?

Suppose our physiologist, whose experience is limited to horses, meets with a zebra for the first time,—will he suppose that this generalisation holds good for zebras also ?

That depends very much on his turn of mind. But we will suppose him to be a bold man. He will say, " The zebra is certainly not a horse, but it is very like one,—so like, that it must be the ' ticket' or mark of a blood-circulation also ; and, I conclude that the zebra has a circulation."

That is a deduction, a very fair deduction, but by no means to be considered scientifically secure. This last quality in fact can only be given by *verification*—that is, by making a zebra the subject of all the experiments performed on the horse. Of course, in the present case, the *deduction* would be *confirmed* by this process of verification, and the result would be, not merely a positive widening of knowledge, but a fair increase of confidence in the truth of one's generalisations in other cases.

Thus, having settled the point in the zebra and horse, our philosopher would have great confidence in the existence of a circulation in the ass. Nay, I fancy most persons would excuse him, if in this case he did not take the trouble to go through the process of verification at all; and it would not be without a parallel in the history of the human mind, if our imaginary physiologist now maintained that he was acquainted with asinine circulation *à priori.*

However, if I might impress any caution upon your minds, it is, the utterly conditional nature of all our knowledge,—the danger of neglecting the process of verification under any circumstances; and the film upon which we rest, the moment our deductions carry us beyond the reach of this great process of verification. There is no better instance of this than is afforded by the history of our knowledge of the circulation of the blood in the animal kingdom until the year 1824. In every animal possessing a circulation at all, which had been observed up to that time, the current of the blood was known to take one definite and invariable direction. Now, there is a class of animals called *Ascidians,* which possess a heart and a circulation, and up to the period of which I speak,

no one would have dreamt of questioning the
propriety of the deduction, that these creatures
have a circulation in one direction ; nor would any
one have thought it worth while to verify the
point.   But, in that year, M. von Hasselt, happening to examine a transparent animal of this class.
found, to his infinite surprise, that after the heart
had beat a certain number of times, it stopped,
and then began beating the opposite way—so as
to reverse the course of the current, which returned
by and by to its original direction.

I have myself timed the heart of these little
animals.   I found it as regular as possible in its
periods of reversal : and I know no spectacle in
the animal kingdom more wonderful than that
which it presents—all the more wonderful that to
this day it remains an unique fact, peculiar to this
class among the whole animated world.   At the
same time I know of no more striking case of the
necessity of the *verification* of even those deductions which seem founded on the widest and
safest inductions.

Such are the methods of Biology—methods
which are obviously identical with those of all
other sciences, and therefore wholly incompetent
to form the ground of any distinction between it
and them.[1]

---

[1] Save for the pleasure of doing so, I need hardly point out
my obligations to Mr. J. S. Mill's *System of Logic*, in this view
of scientific method.

But I shall be asked at once, Do you mean to
say that there is no difference between the habit
of mind of a mathematician and that of a naturalist? Do you imagine that Laplace might have
been put into the Jardin des Plantes, and Cuvier
into the Observatory, with equal advantage to the
progress of the sciences they professed?

To which I would reply, that nothing could be
further from my thoughts. But different habits
and various special tendencies of two sciences do
not imply different methods. The mountaineer
and the man of the plains have very different
habits of progression, and each would be at a loss
in the other's place ; but the method of progression,
by putting one leg before the other, is the same in
each case. Every step of each is a combination of
a lift and a push ; but the mountaineer lifts more
and the lowlander pushes more. And I think the
case of two sciences resembles this.

I do not question for a moment, that while the
Mathematician is busy with deductions *from*
general propositions, the Biologist is more especially occupied with observation, comparison,
and those processes which lead *to* general propositions. All I wish to insist upon is, that this
difference depends not on any fundamental distinction in the sciences themselves, but on the
accidents of their subject-matter, of their relative
complexity, and consequent relative perfection.

The Mathematician deals with two properties of

objects only, number and extension, and all the inductions he wants have been formed and finished ages ago. He is occupied now with nothing but deduction and verification.

The Biologist deals with a vast number of properties of objects, and his inductions will not be completed, I fear, for ages to come; but when they are, his science will be as deductive and as exact as the Mathematics themselves.

Such is the relation of Biology to those sciences which deal with objects having fewer properties than itself. But as the student, in reaching Biology, looks back upon sciences of a less complex and therefore more perfect nature; so, on the other hand, does he look forward to other more complex and less perfect branches of knowledge. Biology deals only with living beings as isolated things—treats only of the life of the individual: but there is a higher division of science still, which considers living beings as aggregates—which deals with the relation of living beings one to another— the science which *observes* men—whose *experiments* are made by nations one upon another, in battle-fields—whose *general propositions* are embodied in history, morality, and religion—whose *deductions* lead to our happiness or our misery—and whose *verifications* so often come too late, and serve only

"To point a moral, or adorn a tale"—

I mean the science of Society or *Sociology*.

I think it is one of the grandest features of Biology, that it occupies this central position in human knowledge. There is no side of the human mind which physiological study leaves uncultivated. Connected by innumerable ties with abstract science, Physiology is yet in the most intimate relation with humanity; and by teaching us that law and order, and a definite scheme of development, regulate even the strangest and wildest manifestations of individual life, she prepares the student to look for a goal even amidst the erratic wanderings of mankind, and to believe that history offers something more than an entertaining chaos —a journal of a toilsome, tragi-comic march nowhither.

The preceding considerations have, I hope, served to indicate the replies which befit the first two of the questions which I set before you at starting, viz. What is the range and position of Physiological Science as a branch of knowledge, and what is its value as a means of mental discipline?

Its *subject-matter* is a large moiety of the universe—its *position* is midway between the physico-chemical and the social sciences. Its *value* as a branch of discipline is partly that which it has in common with all sciences—the training and strengthening of common sense; partly that which is more peculiar to itself—the great exercise which it affords to the faculties of observation and comparison ; and, I may add, the *exactness* of knowledge which it requires on the part of those among its votaries who desire to extend its boundaries.

If what has been said as to the position and scope of Biology be correct, our third question— What is the practical value of physiological instruction ?—might, one would think, be left to answer itself.

On other grounds even, were mankind deserving of the title "rational," which they arrogate to themselves, there can be no question that they would consider, as the most necessary of all branches of instruction for themselves and for their children, that which professes to acquaint them with the conditions of the existence they prize so highly—which teaches them how to avoid disease and to cherish health, in themselves and those who are dear to them.

I am addressing, I imagine, an audience of educated persons ; and yet I dare venture to assert that, with the exception of those of my hearers who may chance to have received a medical education, there is not one who could tell me what is the meaning and use of an act which he performs a score of times every minute, and whose suspension would involve his immediate death ;—I mean the act of breathing—or who could state in precise terms why it is that a confined atmosphere is injurious to health.

The *practical value* of Physiological knowledge !
Why is it that educated men can be found to maintain that a slaughter-house in the midst of a great
city is rather a good thing than otherwise ?—that
mothers persist in exposing the largest possible
amount of surface of their children to the cold, by
the absurd style of dress they adopt, and then
marvel at the peculiar dispensation of Providence,
which removes their infants by bronchitis and
gastric fever ? Why is it that quackery rides
rampant over the land ; and that not long ago, one
of the largest public rooms in this great city could
be filled by an audience gravely listening to the
reverend expositor of the doctrine—that the simple
physiological phænomena known as spirit-rapping,
table-turning, phreno-magnetism, and I know not
what other absurd and inappropriate names, are
due to the direct and personal agency of Satan ?

Why is all this, except from the utter ignorance
as to the simplest laws of their own animal life,
which prevails among even the most highly educated persons in this country ?

But there are other branches of Biological
Science, besides Physiology proper, whose practical
influence, though less obvious, is not, as I believe,
less certain.   I have heard educated men speak
with an ill-disguised contempt of the studies of the
naturalist, and ask, not without a shrug, " What is
the use of knowing all about these miserable
animals—what bearing has it on human life ? "

I will endeavour to answer that question. I take it that all will admit there is definite Government of this universe—that its pleasures and pains are not scattered at random, but are distributed in accordance with orderly and fixed laws, and that it is only in accordance with all we know of the rest of the world, that there should be an agreement between one portion of the sensitive creation and another in these matters.

Surely then it interests us to know the lot of other animal creatures—however far below us, they are still the sole created things which share with us the capability of pleasure and the susceptibility to pain.

I cannot but think that he who finds a certain proportion of pain and evil inseparably woven up in the life of the very worms, will bear his own share with more courage and submission; and will, at any rate, view with suspicion those weakly amiable theories of the Divine government, which would have us believe pain to be an oversight and a mistake,—to be corrected by and by. On the other hand, the predominance of happiness among living things—their lavish beauty—the secret and wonderful harmony which pervades them all, from the highest to the lowest, are equally striking refutations of that modern Manichean doctrine, which exhibits the world as a slave-mill, worked with many tears, for mere utilitarian ends.

There is yet another way in which natural history

may, I am convinced, take a profound hold upon
practical life,—and that is, by its influence over
our finer feelings, as the greatest of all sources of
that pleasure which is derivable from beauty.   I
do not pretend that natural-history knowledge, as
such, can increase our sense of the beautiful in
natural objects.   I do not suppose that the dead
soul of Peter Bell, of whom the great poet of
nature says,—

> A primrose by the river s brim,
> A yellow primrose was to him,—
> And it was nothing more,—

would have been a whit roused from its apathy by
the information that the primrose is a Dicotyle-
donous Exogen, with a monopetalous corolla and
central placentation.   But I advocate natural-
history knowledge from this point of view, because
it would lead us to *seek* the beauties of natural
objects, instead of trusting to chance to force them
on our attention.   To a person uninstructed in
natural history, his country or sea-side stroll is a
walk through a gallery filled with wonderful works
of art, nine-tenths of which have their faces turned
to the wall.   Teach him something of natural
history, and you place in his hands a catalogue of
those which are worth turning round.   Surely our
innocent pleasures are not so abundant in this life,
that we can afford to despise this or any other
source of them.   We should fear being banished
for our neglect to that limbo, where the great

Florentine tells us are those who, during this life, " wept when they might be joyful."

But I shall be trespassing unwarrantably on your kindness, if I do not proceed at once to my last point—the time at which Physiological Science should first form a part of the Curriculum of Education.

The distinction between the teaching of the facts of a science as instruction, and the teaching it systematically as knowledge, has already been placed before you in a previous lecture : and it appears to me that, as with other sciences, the *common facts* of Biology—the uses of parts of the body—the names and habits of the living creatures which surround us—may be taught with advantage to the youngest child. Indeed, the avidity of children for this kind of knowledge, and the comparative ease with which they retain it, is something quite marvellous. I doubt whether any toy would be so acceptable to young children as a vivarium of the same kind as, but of course on a smaller scale than, those admirable devices in the Zoological Gardens.

On the other hand, systematic teaching in Biology cannot be attempted with success until the student has attained to a certain knowledge of physics and chemistry : for though the phæ-nomena of life are dependent neither on physical nor on chemical, but on vital forces, yet they result in all sorts of physical and chemical changes, which can only be judged by their own laws.

And now to sum up in a few words the conclusions to which I hope you see reason to follow me.

Biology needs no apologist when she demands a place—and a prominent place—in any scheme of education worthy of the name. Leave out the Physiological sciences from your curriculum, and you launch the student into the world, undisciplined in that science whose subject-matter would best develop his powers of observation; ignorant of facts of the deepest importance for his own and others' welfare; blind to the richest sources of beauty in God's creation; and unprovided with that belief in a living law, and an order manifesting itself in and through endless change and variety, which might serve to check and moderate that phase of despair through which, if he take an earnest interest in social problems, he will assuredly sooner or later pass.

Finally, one word for myself. I have not hesitated to speak strongly where I have felt strongly; and I am but too conscious that the indicative and imperative moods have too often taken the place of the more becoming subjunctive and conditional. I feel, therefore, how necessary it is to beg you to forget the personality of him who has thus ventured to address you, and to consider only the truth or error in what has been said.

III

EMANCIPATION—BLACK AND WHITE

[1865.]

QUASHIE'S plaintive inquiry, "Am I not a man and a brother?" seems at last to have received its final reply—the recent decision of the fierce trial by battle on the other side of the Atlantic fully concurring with that long since delivered here in a more peaceful way.

The question is settled; but even those who are most thoroughly convinced that the doom is just, must see good grounds for repudiating half the arguments which have been employed by the winning side; and for doubting whether its ultimate results will embody the hopes of the victors, though they may more than realise the fears of the vanquished. It may be quite true that some negroes are better than some white men; but no rational man, cognisant of the facts, believes that the average negro is the equal, still less the superior, of the average white man. And, if this be true, it is simply incredible that, when all his disabilities are removed, and our prognathous relative has a fair field and no favour, as well as no oppressor, he will be able to compete successfully with his bigger-brained and smaller-jawed rival, in a contest which is to be carried on by thoughts and not by bites. The highest places in the hierarchy of civilisation will assuredly not be within the reach of our dusky cousins, though it is by no means necessary that they should be restricted to the lowest. But whatever the position of stable equilibrium into which the laws of social gravitation may bring the negro, all responsibility for the result will henceforward lie between Nature and him. The white man may wash his hands of it, and the Caucasian conscience be void of reproach for evermore. And this, if we look to the bottom of the matter, is the real justification for the abolition policy.

The doctrine of equal natural rights may be an illogical delusion ; emancipation may convert the slave from a well-fed animal into a pauperised man ; mankind may even have to do without cotton shirts ; but all these evils must be faced if the moral law, that no human being can arbitrarily dominate over another without grievous damage to his own nature, be, as many think, as readily demonstrable by experiment as any physical truth. If this be true, no slavery can

F 2

be abolished without a double emancipation, and the master will benefit by freedom more than the freed-man.

The like considerations apply to all the other questions of emancipation which are at present stirring the world—the multifarious demands that classes of mankind shall be relieved from restrictions imposed by the artifice of man, and not by the necessities of Nature.   One of the most important, if not the most important, of all these, is that which daily threatens to become the "irrepressible" woman question.   What social and political rights have women ?   What ought they to be allowed, or not allowed, to do, be, and suffer ?   And, as involved in, and underlying all these questions, how ought they to be educated ?

There are philogynists as fanatical as any "misogynists" who, reversing our antiquated notions, bid the man look upon the woman as the higher type of humanity; who ask us to regard the female intellect as the clearer and the quicker, if not the stronger; who desire us to look up to the feminine moral sense as the purer and the nobler; and bid man abdicate his usurped sovereignty over Nature in favour of the female line.   On the other hand, there are persons not to be outdone in all loyalty and just respect for womankind, but by nature hard of head and haters of delusion, however charming, who not only repudiate the new woman-worship which so many sentimentalists and some philosophers are desirous of setting up, but, carrying their audacity further, deny even the natural equality of the sexes. They assert, on the contrary, that in every excellent character, whether mental or physical, the average woman is inferior to the average man, in the sense of having that character less in quantity and lower in quality. Tell these persons of the rapid perceptions and the instinctive intellectual insight of women, and they reply that the feminine mental peculiarities, which pass under these names, are merely the outcome of a greater impressibility to the superficial aspects of things, and of the absence of that restraint upon expression which, in men, is imposed by reflection and a sense of responsibility. Talk of the passive endurance of the weaker sex, and opponents of this kind remind you that Job was a man, and that, until quite recent times, patience and long-suffering were not counted among the specially feminine virtues. Claim passionate tenderness as especially feminine, and the inquiry is made whether all the best love-poetry in existence (except, perhaps, the "Sonnets from the Portuguese") has not been written by men; whether the song which embodies the ideal of pure and tender passion—"Adelaida" —was written by *Frau* Beethoven; whether it was the Fornarina, or Raphael, who painted the Sistine Madonna. Nay, we have known one such

sensical. And we conceive that those who may laugh at the arguments of the extreme philogynists, may yet feel bound to work heart and soul towards the attainment of their practical ends.

As regards education, for example. Granting the alleged defects of women, is it not somewhat absurd to sanction and maintain a system of education which would seem to have been specially contrived to exaggerate all these defects?

Naturally not so firmly strung, nor so well balanced as boys, girls are in great measure debarred from the sports and physical exercises which are justly thought absolutely necessary for the full development of the vigour of the more favoured sex. Women are, by nature, more excitable than men—prone to be swept by tides of emotion, proceeding from hidden and inward, as well as from obvious and external causes; and female education does its best to weaken every physical counterpoise to this nervous mobility—tends in all ways to stimulate the emotional part of the mind and stunt the rest. We find girls naturally timid, inclined to dependence, born conservatives; and we teach them that independence is unladylike; that blind faith is the right frame of mind; and that whatever we may be permitted, and indeed encouraged, to do to our brother, our sister is to be left to the tyranny of authority and tradition. With few insignificant exceptions, girls have been educated either to be drudges, or toys, beneath man; or a sort of angels above him; the highest ideal aimed at oscillating between Clärchen and Beatrice. The possibility that the ideal of womanhood lies neither in the fair saint, nor in the fair sinner; that the female type of character is neither better nor worse than the male, but only weaker; that women are meant neither to be men's guides nor their playthings, but their comrades, their fellows, and their equals, so far as Nature puts no bar to that equality, does not seem to have entered into the minds of those who have had the conduct of the education of girls.

If the present system of female education stands self-condemned, as inherently absurd; and if that which we have just indicated is the true position of woman, what is the first step towards a better state of things? We reply, emancipate girls. Recognise the fact that they share the senses, perceptions, feelings, reasoning powers, emotions, of boys, and that the mind of the average girl is less different from that of the average boy, than the mind of one boy is from that of another; so that whatever argument justifies a given education for all boys, justifies its application to girls as well. So far from imposing artificial restrictions upon the acquirement of knowledge by women, throw every facility in their

way.   Let our Faustinas, if they will, toil through
the whole round of

> " Juristerei und Medizin,
> Und leider ! auch Philosophie."

Let us have " sweet girl graduates " by all means.
They will be none the less sweet for a little
wisdom ; and the " golden hair " will not curl less
gracefully outside the head by reason of there
being brains within.   Nay, if obvious practical
difficulties can be overcome, let those women who
feel inclined to do so descend into the gladiatorial
arena of life, not merely in the guise of *retiariæ*,
as heretofore, but as bold *sicariæ*, breasting the
open fray.   Let them, if they so please, become
merchants, barristers, politicians.   Let them have
a fair field, but let them understand, as the
necessary correlative, that they are to have no
favour.   Let Nature alone sit high above the lists,
" rain influence and judge the prize."

And the result ?   For our parts, though loth to
prophesy, we believe it will be that of other
emancipations.   Women will find their place, and
it will neither be that in which they have been
held, nor that to which some of them aspire.
Nature's old salique law will not be repealed, and
no change of dynasty will be effected.   The big
chests, the massive brains, the vigorous muscles
and stout frames of the best men will carry the
day, whenever it is worth their while to contest

the prizes of life with the best women. And the
hardship of it is, that the very improvement of
the women will lessen their chances. Better
mothers will bring forth better sons, and the
impetus gained by the one sex will be transmitted,
in the next generation, to the other. The most
Darwinian of theorists will not venture to propound the doctrine, that the physical disabilities
under which women have hitherto laboured in
the struggle for existence with men are likely to
be removed by even the most skilfully conducted
process of educational selection.

We are, indeed, fully prepared to believe that
the bearing of children may, and ought, to become
as free from danger and long disability to the
civilised woman as it is to the savage ; nor is it
improbable that, as society advances towards its
right organisation, motherhood will occupy a less
space of woman's life than it has hitherto done.
But still, unless the human species is to come to
an end altogether—a consummation which can
hardly be desired by even the most ardent advocate of "women's rights"—somebody must be
good enough to take the trouble and responsibility
of annually adding to the world exactly as many
people as die out of it. In consequence of some
domestic difficulties, Sydney Smith is said to
have suggested that it would have been good for
the human race had the model offered by the hive
been followed, and had all the working part of the

female community been neuters. Failing any thorough-going reform of this kind, we see nothing for it but the old division of humanity into men potentially, or actually, fathers, and women potentially, if not actually, mothers. And we fear that so long as this potential motherhood is her lot, woman will be found to be fearfully weighted in the race of life.

The duty of man is to see that not a grain is piled upon that load beyond what Nature imposes; that injustice is not added to inequality.

A LIBERAL EDUCATION; AND WHERE TO FIND IT

[1868.]

THE business which the South London Working Men's College has undertaken is a great work; indeed, I might say, that Education, with which that college proposes to grapple, is the greatest work of all those which lie ready to a man's hand just at present.

And, at length, this fact is becoming generally recognised. You cannot go anywhere without hearing a buzz of more or less confused and contradictory talk on this subject—nor can you fail to notice that, in one point at any rate, there is a very decided advance upon like discussions in former days. Nobody outside the agricultural interest now dares to say that education is a bad thing. If any representative of the once large and powerful party, which, in former days, proclaimed this opinion, still exists in a semi-fossil state, he keeps his thoughts to himself. In fact, there is a chorus of voices, almost distressing in their harmony, raised in favour of the doctrine that education is the great panacea for human troubles, and that, if the country is not shortly to go to the dogs, everybody must be educated.

The politicians tells us, " You must educate the masses because they are going to be masters." The clergy join in the cry for education, for they affirm that the people are drifting away from church and chapel into the broadest infidelity. The manufacturers and the capitalists swell the chorus lustily. They declare that ignorance makes bad workmen; that England will soon be unable to turn out cotton goods, or steam engines, cheaper than other people; and then, Ichabod ! Ichabod ! the glory will be departed from us. And a few voices are lifted up in favour of the doctrine that the masses should be educated because they are men and women with unlimited capacities of being, doing, and suffering, and that it us as true now, as ever it was, that the people perish for lack of knowledge.

These members of the minority, with whom I confess I have a good deal of sympathy, are doubtful whether any of the other reasons urged in favour of the education of the people are of much value—whether, indeed, some of them are based upon either wise or noble grounds of action. They question if it be wise to tell people that you

will do for them, out of fear of their power, what
you have left undone, so long as your only motive
was compassion for their weakness and their sor-
rows. And, if ignorance of everything which it
is needful a ruler should know is likely to do so
much harm in the governing classes of the future,
why is it, they ask reasonably enough, that such
ignorance in the governing classes of the past has
not been viewed with equal horror ?

Compare the average artisan and the average
country squire, and it may be doubted if you will
find a pin to choose between the two in point of
ignorance, class feeling, or prejudice. It is true
that the ignorance is of a different sort—that the
class feeling is in favour of a different class—
and that the prejudice has a distinct savour of
wrong-headedness in each case—but it is questionable if the one is either a bit better, or a bit worse,
than the other. The old protectionist theory is
the doctrine of trades unions as applied by the
squires, and the modern trades unionism is the
doctrine of the squires applied by the artisans.
Why should we be worse off under one *régime* than
under the other ?

Again, this sceptical minority asks the clergy to
think whether it is really want of education which
keeps the masses away from their ministrations—
whether the most completely educated men are
not as open to reproach on this score as the workmen ; and whether, perchance, this may not indi-

cate that it is not education which lies at the bottom of the matter?

Once more, these people, whom there is no pleasing, venture to doubt whether the glory, which rests upon being able to undersell all the rest of the world, is a very safe kind of glory—whether we may not purchase it too dear; especially if we allow education, which ought to be directed to the making of men, to be diverted into a process of manufacturing human tools, wonderfully adroit in the exercise of some technical industry, but good for nothing else.

And, finally, these people inquire whether it is the masses alone who need a reformed and improved education. They ask whether the richest of our public schools might not well be made to supply knowledge, as well as gentlemanly habits, a strong class feeling, and eminent proficiency in cricket. They seem to think that the noble foundations of our old universities are hardly fulfilling their functions in their present posture of half-clerical seminaries, half racecourses, where men are trained to win a senior wranglership, or a double-first, as horses are trained to win a cup, with as little reference to the needs of after-life in the case of the man as in that of the racer. And, while as zealous for education as the rest, they affirm that, if the education of the richer classes were such as to fit them to be the leaders and the governors of the poorer; and, if the education of the

poorer classes were such as to enable them to appreciate really wise guidance and good governance, the
politicians need not fear mob-law, nor the clergy
lament their want of flocks, nor the capitalists prog-
nosticate the annihilation of the prosperity of the
country.

Such is the diversity of opinion upon the why
and the wherefore of education. And my hearers
will be prepared to expect that the practical recommendations which are put forward are not less
discordant. There is a loud cry for compulsory
education. We English, in spite of constant experience to the contrary, preserve a touching faith
in the efficacy of acts of Parliament; and I believe
we should have compulsory education in the course
of next session, if there were the least probability
that half a dozen leading statesmen of different
parties would agree what that education should be.

Some hold that education without theology is
worse than none. Others maintain, quite as
strongly, that education with theology is in the
same predicament. But this is certain, that those
who hold the first opinion can by no means agree
what theology should be taught; and that those
who maintain the second are in a small minority.

At any rate "make people learn to read, write,
and cipher," say a great many; and the advice is
undoubtedly sensible as far as it goes. But, as
has happened to me in former days, those who, in
despair of getting anything better, advocate this

measure, are met with the objection that it is very like making a child practise the use of a knife, fork, and spoon, without giving it a particle of meat. I really don't know what reply is to be made to such an objection.

But it would be unprofitable to spend more time in disentangling, or rather in showing up the knots in, the ravelled skeins of our neighbours. Much more to the purpose is it to ask if we possess any clue of our own which may guide us among these entanglements. And by way of a beginning, let us ask ourselves—What is education ? Above all things, what is our ideal of a thoroughly liberal education ?—of that education which, if we could begin life again, we would give ourselves— of that education which, if we could mould the fates to our own will, we would give our children ? Well, I know not what may be your conceptions upon this matter, but I will tell you mine, and I hope I shall find that our views are not very discrepant.

Suppose it were perfectly certain that the life and fortune of every one of us would, one day or other, depend upon his winning or losing a game at chess. Don't you think that we should all consider it to be a primary duty to learn at least the names and the moves of the pieces ; to have a notion of a gambit, and a keen eye for all the means of giving and getting out of check ? Do

you not think that we should look with a disap
probation amounting to scorn, upon the father who
allowed his son, or the state which allowed its
members, to grow up without knowing a pawn
from a knight ?

Yet it is a very plain and elementary truth, that
the life, the fortune, and the happiness of every
one of us, and, more or less, of those who are connected with us, do depend upon our knowing
something of the rules of a game infinitely more
difficult and complicated than chess.   It is a game
which has been played for untold ages, every man
and woman of us being one of the two players in a
game of his or her own.   The chess-board is the
world, the pieces are the phenomena of the
universe, the rules of the game are what we call
the laws of Nature.   The player on the other side
is hidden from us.   We know that his play is
always fair, just and patient.   But also we know,
to our cost, that he never overlooks a mistake, or
makes the smallest allowance for ignorance.   To
the man who plays well, the highest stakes are
paid, with that sort of overflowing generosity with
which the strong shows delight in strength.   And
one who plays ill is checkmated—without haste,
but without remorse.

My metaphor will remind some of you
of the famous picture in which Retzsch has
depicted Satan playing at chess with man for his
soul. Substitute for the mocking fiend in that

picture a calm, strong angel who is playing for love, as we say, and would rather lose than win—and I should accept it as an image of human life.

Well, what I mean by Education is learning the rules of this mighty game. In other words, education is the instruction of the intellect in the laws of Nature, under which name I include not merely things and their forces, but men and their ways; and the fashioning of the affections and of the will into an earnest and loving desire to move in harmony with those laws. For me, education means neither more nor less than this. Anything which professes to call itself education must be tried by this standard, and if it fails to stand the test, I will not call it education, whatever may be the force of authority, or of numbers, upon the other side.

It is important to remember that, in strictness, there is no such thing as an uneducated man. Take an extreme case. Suppose that an adult man, in the full vigour of his faculties, could be suddenly placed in the world, as Adam is said to have been, and then left to do as he best might. How long would he be left uneducated? Not five minutes. Nature would begin to teach him, through the eye, the ear, the touch, the properties of objects. Pain and pleasure would be at his elbow telling him to do this and avoid that; and by slow degrees the man would receive an education which, if narrow, would be thorough, real, and adequate to his circumstances, though there would be no extras and very few accomplishments.

And if to this solitary man entered a second Adam, or, better still, an Eve, a new and greater world, that of social and moral phenomena, would be revealed. Joys and woes, compared with which all others might seem but faint shadows, would spring from the new relations. Happiness and sorrow would take the place of the coarser monitors, pleasure and pain; but conduct would still be shaped by the observation of the natural consequences of actions; or, in other words, by the laws of the nature of man.

To every one of us the world was once as fresh and new as to Adam. And then, long before we were susceptible of any other mode of instruction, Nature took us in hand, and every minute of waking life brought its educational influence, shaping our actions into rough accordance with Nature's laws, so that we might not be ended untimely by too gross disobedience. Nor should I speak of this process of education as past for any one, be he as old as he may. For every man the world is as fresh as it was at the first day, and as full of untold novelties for him who has the eyes to see them. And Nature is still continuing her patient education of us in that great university, the universe, of which we are all members—Nature having no Test-Acts.

Those who take honours in Nature's university, who learn the laws which govern men and things and obey them, are the really great and successful men in this world. The great mass of mankind are the " Poll," who pick up just enough to get through without much discredit. Those who won't learn at all are plucked ; and then you can't come up again. Nature's pluck means extermination.

Thus the question of compulsory education is settled so far as Nature is concerned. Her bill on that question was framed and passed long ago. But, like all compulsory legislation, that of Nature is harsh and wasteful in its operation. Ignorance is visited as sharply as wilful disobedience—incapacity meets with the same punishment as crime. Nature's discipline is not even a word and a blow, and the blow first ; but the blow without the word. It is left to you to find out why your ears are boxed.

The object of what we commonly call education—that education in which man intervenes and which I shall distinguish as artificial education—is to make good these defects in Nature's methods; to prepare the child to receive Nature's education, neither incapably nor ignorantly, nor with wilful disobedience; and to understand the preliminary symptoms of her pleasure, without waiting for the box on the ear. In short, all artificial education ought to be an anticipation of natural education. And a liberal education is an artificial education which has not only prepared a man to escape the great evils of disobedience to natural laws, but has trained him to appreciate and to seize upon the rewards, which Nature scatters with as free a hand as her penalties.

That man, I think, has had a liberal education who has been so trained in youth that his body is the ready servant of his will, and does with ease and pleasure all the work that, as a mechanism, it is capable of; whose intellect is a clear, cold, logic engine, with all its parts of equal strength, and in smooth working order; ready, like a steam engine, to be turned to any kind of work, and spin the gossamers as well as forge the anchors of the mind; whose mind is stored with a knowledge of the great and fundamental truths of Nature and of the laws of her operations; one who, no stunted ascetic, is full of life and fire, but whose passions are trained to come to heel by a vigorous will, the servant of a tender conscience; who has learned to love all beauty, whether of Nature or of art, to hate all vileness, and to respect others as himself.

Such an one and no other, I conceive, has had a liberal education; for he is, as completely as a man can be, in harmony with Nature. He will make the best of her, and she of him. They will get on together rarely : she as his ever beneficent mother; he as her mouthpiece, her conscious self, her minister and interpreter.

Where is such an education as this to be had ?

Where is there any approximation to it? Has any one tried to found such an education? Looking over the length and breadth of these islands, I am afraid that all these questions must receive a negative answer. Consider our primary schools and what is taught in them. A child learns:—

1. To read, write, and cipher, more or less well; but in a very large proportion of cases not so well as to take pleasure in reading, or to be able to write the commonest letter properly.

2. A quantity of dogmatic theology, of which the child, nine times out of ten, understands next to nothing.

3. Mixed up with this, so as to seem to stand or fall with it, a few of the broadest and simplest principles of morality. This, to my mind, is much as if a man of science should make the story of the fall of the apple in Newton's garden an integral part of the doctrine of gravitation, and teach it as of equal authority with the law of the inverse squares.

4. A good deal of Jewish history and Syrian geography, and perhaps a little something about English history and the geography of the child's own country. But I doubt if there is a primary school in England in which hangs a map of the hundred in which the village lies, so that the children may be practically taught by it what a map means.

5. A certain amount of regularity, attentive obedience, respect for others: obtained by fear, if the master be incompetent or foolish ; by love and reverence, if he be wise.

So far as this school course embraces a training in the theory and practice of obedience to the moral laws of Nature, I gladly admit, not only that it contains a valuable educational element, but that, so far, it deals with the most valuable and important part of all education. Yet, contrast what is done in this direction with what might be done; with the time given to matters of comparatively no importance; with the absence of any attention to things of the highest moment; and one is tempted to think of Falstaff's bill and " the halfpenny worth of bread to all that quantity of sack."

Let us consider what a child thus " educated " knows, and what it does not know. Begin with the most important topic of all—morality, as the guide of conduct. The child knows well enough that some acts meet with approbation and some with disapprobation. But it has never heard that there lies in the nature of things a reason for every moral law, as cogent and as well defined as that which underlies every physical law ; that stealing and lying are just as certain to be followed by evil consequences, as putting your hand in the fire, or jumping out of a garret window. Again, though the scholar may have been made acquainted, in dogmatic fashion, with the broad laws of morality, he has had no training in the application of those laws to the difficult problems which result from the complex conditions of modern civilisation. Would it not be very hard to expect any one to solve a problem in conic sections who had merely been taught the axioms and definitions of mathematical science?

A workman has to bear hard labour, and perhaps privation, while he sees others rolling in wealth, and feeding their dogs with what would keep his children from starvation. Would it not be well to have helped that man to calm the natural promptings of discontent by showing him, in his youth, the necessary connection of the moral law which prohibits stealing with the stability of society—by proving to him, once for all, that it is better for his own people, better for himself, better for future generations, that he should starve than steal? If you have no foundation of knowledge, or habit of thought, to work upon, what chance have you of persuading a hungry man that a capitalist is not a thief "with a circumbendibus?" And if he honestly believes that, of what avail is it to quote the commandment against stealing, when he proposes to make the capitalist disgorge?

Again, the child learns absolutely nothing of the history or the political organisation of his own country. His general impression is, that everything of much importance happened a very long while ago; and that the Queen and the gentlefolks govern the country much after the fashion of King David and the elders and nobles of Israel—his sole models. Will you give a man with this much information a vote ? In easy times he sells it for a pot of beer. Why should he not? It is of about as much use to him as a chignon, and he knows as much what to do with it, for any other purpose. In bad times, on the contrary, he applies his simple theory of government, and believes that his rulers are the cause of his sufferings—a belief which sometimes bears remarkable practical fruits.

Least of all, does the child gather from this primary "education" of ours a conception of the laws of the physical world, or of the relations of cause and effect therein. And this is the more to be lamented, as the poor are especially exposed to physical evils, and are more interested in removing them than any other class of the community. If any one is concerned in knowing the ordinary laws of mechanics one would think it is the hand-labourer, whose daily toil lies among levers and pulleys; or among the other implements of artisan work. And if any one is interested in the laws of health, it is the poor workman, whose strength is wasted by ill-prepared food, whose health is sapped by bad ventilation and bad drainage, and half whose children are massacred by disorders which might be prevented. Not only does our present primary education carefully abstain from hinting to the workman that some of his greatest evils are traceable to mere physical agencies, which could be removed by energy, patience, and frugality; but it does worse—it renders him, so far as it can, deaf to those who could help him, and tries to substitute an Oriental submission to what is falsely declared to be the will of God, for his natural tendency to strive after a better condition.

What wonder, then, if very recently an appeal has been made to statistics for the profoundly foolish purpose of showing that education is of no good —that it diminishes neither misery nor crime among the masses of mankind? I reply, why should the thing which has been called education do either the one or the other? If I am a knave or a fool, teaching me to read and write won't make me less of either one or the other—unless somebody shows me how to put my reading and writing to wise and good purposes.

Suppose any one were to argue that medicine is of no use, because it could be proved statistically, that the percentage of deaths was just the same among people who had been taught how to open a medicine chest, and among those who did not so much as know the key by sight. The argument is absurd; but it is not more preposterous than that against which I am contending. The only medicine for suffering, crime, and all the other woes of mankind, is wisdom. Teach a man to read and write, and you have put into his hands the great keys of the wisdom box. But it is quite another matter whether he ever opens the box or not. And he is as likely to poison as to cure himself, if, without guidance, he swallows the first drug that comes to hand. In these times a man may as well be purblind, as unable to read— lame, as unable to write. But I protest that, if I thought the alternative were a necessary one, I would rather that the children of the poor should grow up ignorant of both these mighty arts, than that they should remain ignorant of that know- ledge to which these arts are means.

It may be said that all these animadversions may apply to primary schools, but that the higher schools, at any rate, must be allowed to give a liberal education. In fact they professedly sacrifice everything else to this object.

Let us inquire into this matter. What do the higher schools, those to which the great middle class of the country sends its children, teach, over and above the instruction given in the primary schools? There is a little more reading and writing of English. But, for all that, every one knows that it is a rare thing to find a boy of the middle or upper classes who can read aloud decently, or who can put his thoughts on paper in clear and gram- matical (to say nothing of good or elegant) language.

The "ciphering" of the lower schools expands into elementary mathematics in the higher; into arithmetic, with a little algebra, a little Euclid. But I doubt if one boy in five hundred has ever heard the explanation of a rule of arithmetic, or knows his Euclid otherwise than by rote.

Of theology, the middle class schoolboy gets rather less than poorer children, less absolutely and less relatively, because there are so many other claims upon his attention. I venture to say that, in the great majority of cases, his ideas on this subject when he leaves school are of the most shadowy and vague description, and associated with painful impressions of the weary hours spent in learning collects and catechism by heart.

Modern geography, modern history, modern literature; the English language as a language; the whole circle of the sciences, physical, moral and social, are even more completely ignored in the higher than in the lower schools. Up till within a few years back, a boy might have passed through any one of the great public schools with the greatest distinction and credit, and might never so much as have heard of one of the subjects I have just mentioned. He might never have heard that the earth goes round the sun; that England underwent a great revolution in 1688, and France another in 1789; that there once lived certain notable men called Chaucer, Shakespeare, Milton, Voltaire, Goethe, Schiller. The first might be a German and the last an Englishman for anything he could tell you to the contrary. And as for Science, the only idea the word would suggest to his mind would be dexterity in boxing.

I have said that this was the state of things a few years back, for the sake of the few righteous who are to be found among the educational cities of the plain. But I would not have you too sanguine about the result, if you sound the minds of the existing generation of public schoolboys, on such topics as those I have mentioned.

Now let us pause to consider this wonderful state of affairs; for the time will come when Englishmen will quote it as the stock example of the stolid stupidity of their ancestors in the nineteenth century. The most thoroughly commercial people, the greatest voluntary wanderers and colonists the world has ever seen, are precisely the middle classes of this country. If there be a people which has been busy making history on the great scale for the last three hundred years—and the most profoundly interesting history—history which, if it happened to be that of Greece or Rome, we should study with avidity—it is the English. If there be a people which, during the same period, has developed a remarkable literature, it is our own. If there be a nation whose prosperity depends absolutely and wholly upon their mastery over the forces of Nature, upon their intelligent apprehension of, and obedience to the laws of the creation and distribution of wealth, and of the stable equilibrium of the forces of society, it is precisely this nation. And yet this is what these wonderful people tell their sons :—" At the cost of from one to two thousand pounds of our hard-earned money, we devote twelve of the most precious years of your lives to school. There you shall toil, or be supposed to toil ; but there you shall not learn one single thing of all those you will most want to know directly you leave school and enter upon the practical business of life. You will in all probability go into business, but you shall not know where, or how, any article of commerce is produced, or the difference between an export or an import, or the meaning of the word " capital." You will very likely settle in a colony, but you shall not know whether Tasmania is part of New South Wales, or *vice versâ*.

" Very probably you may become a manufacturer, but you shall not be provided with the means of understanding the working of one of your own steam-engines, or the nature of the raw products you employ ; and, when you are asked to buy a patent, you shall not have the slightest means of judging whether the inventor is an impostor who is contravening the elementary principles of science, or a man who will make you as rich as Crœsus.

" You will very likely get into the House of

Commons. You will have to take your share in making laws which may prove a blessing or a curse to millions of men. But you shall not hear one word respecting the political organisation of your country; the meaning of the controversy between free-traders and protectionists shall never have been mentioned to you ; you shall not so much as know that there are such things as economical laws.

" The mental power which will be of most importance in your daily life will be the power of seeing things as they are without regard to authority; and of drawing accurate general conclusions from particular facts. But at school and at college you shall know of no source of truth but authority; nor exercise your reasoning faculty upon anything but deduction from that which is laid down by authority.

" You will have to weary your soul with work, and many a time eat your bread in sorrow and in bitterness, and you shall not have learned to take refuge in the great source of pleasure without alloy, the serene resting-place for worn human nature,—the world of art."

Said I not rightly that we are a wonderful people ? I am quite prepared to allow, that education entirely devoted to these omitted subjects might not be a completely liberal education. But is an education which ignores them all a liberal education ? Nay, is it too much to say that the education which should embrace these subjects and no others would be a real education, though an incomplete one; while an education which omits them is really not an education at all, but a more or less useful course of intellectual gymnastics?

For what does the middle-class school put in the place of all these things which are left out? It substitutes what is usually comprised under the compendious title of the "classics"—that is to say, the languages, the literature, and the history of the ancient Greeks and Romans, and the geography of so much of the world as was known to these two great nations of antiquity. Now, do not expect me to depreciate the earnest and enlightened pursuit of classical learning. I have not the least desire to speak ill of such occupations, nor any sympathy with those who run them down. On the contrary, if my opportunities had lain in that direction, there is no investigation into which I could have thrown myself with greater delight than that of antiquity.

What science can present greater attractions than philology? How can a lover of literary excellence fail to rejoice in the ancient master-pieces? And with what consistency could I, whose business lies so much in the attempt to de-cipher the past, and to build up intelligible forms out of the scattered fragments of long-extinct beings, fail to take a sympathetic, though an

unlearned, interest in the labours of a Niebuhr, a
Gibbon, or a Grote ?   Classical history is a great
section of the palæontology of man ; and I have
the same double respect for it as for other kinds
of palæontology—that is to say, a respect for the
facts which it establishes as for all facts, and a
still greater respect for it as a preparation for the
discovery of a law of progress.

But if the classics were taught as they might be
taught—if boys and girls were instructed in Greek
and Latin, not merely as languages, but as illus-
trations of philological science ; if a vivid picture
of life on the shores of the Mediterranean two
thousand years ago were imprinted on the minds
of scholars ; if ancient history were taught, not as
a weary series of feuds and fights, but traced to its
causes in such men placed under such conditions ;
if, lastly, the study of the classical books were
followed in such a manner as to impress boys with
their beauties, and with the grand simplicity of
their statement of the everlasting problems of
human life, instead of with their verbal and grammatical peculiarities ; I still think it as little
proper that they should form the basis of a liberal
education for our contemporaries, as I should
think it fitting to make that sort of palæontology
with which I am familiar the back-bone of
modern education.

It is wonderful how close a parallel to classical
training could be made out of that palæontology

to which I refer. In the first place I could get up an osteological primer so arid, so pedantic in its terminology, so altogether distasteful to the youthful mind, as to beat the recent famous production of the head-masters out of the field in all these excellences. Next, I could exercise my boys upon easy fossils, and bring out all their powers of memory and all their ingenuity in the application of my osteo-grammatical rules to the interpretation, or construing, of those fragments. To those who had reached the higher classes, I might supply odd bones to be built up into animals, giving great honour and reward to him who succeeded in fabricating monsters most entirely in accordance with the rules. That would answer to verse-making and essay-writing in the dead languages.

To be sure, if a great comparative anatomist were to look at these fabrications he might shake his head, or laugh. But what then? Would such a catastrophe destroy the parallel? What, think you, would Cicero, or Horace, say to the production of the best sixth form going? And would not Terence stop his ears and run out if he could be present at an English performance of his own plays? Would *Hamlet*, in the mouths of a set of French actors, who should insist on pronouncing English after the fashion of their own tongue, be more hideously ridiculous?

But it will be said that I am forgetting the beauty, and the human interest, which appertain

to classical studies. To this I reply that it is only
a very strong man who can appreciate the charms
of a landscape as he is toiling up a steep hill,
along a bad road. What with short-windedness,
stones, ruts, and a pervading sense of the wisdom
of rest and be thankful, most of us have little
enough sense of the beautiful under these circumstances. The ordinary schoolboy is precisely in
this case. He finds Parnassus uncommonly steep,
and there is no chance of his having much time or
inclination to look about him till he gets to the
top. And nine times out of ten he does not get to
the top.

  But if this be a fair picture of the results of classical teaching at its best—and I gather from those who
have authority to speak on such matters that it is so
—what is to be said of classical teaching at its worst,
or in other words, of the classics of our ordinary
middle-class schools ? [1] I will tell you. It means
getting up endless forms and rules by heart. It
means turning Latin and Greek into English, for
the mere sake of being able to do it, and without
the smallest regard to the worth, or worthlessness,
of the author read. It means the learning of innumerable, not always decent, fables in such a
shape that the meaning they once had is dried up
into utter trash ; and the only impression left upon
a boy's mind is, that the people who believed such

  [1] For a justification of what is here said about these schools,
see that valuable book, *Essays on a Liberal Education, passim,*

things must have been the greatest idiots the world ever saw. And it means, finally, that after a dozen years spent at this kind of work, the sufferer shall be incompetent to interpret a passage in an author he has not already got up; that he shall loathe the sight of a Greek or Latin book; and that he shall never open, or think of, a classical writer again, until, wonderful to relate, he insists upon submitting his sons to the same process.

These be your gods, O Israel! For the sake of this net result (and respectability) the British father denies his children all the knowledge they might turn to account in life, not merely for the achievement of vulgar success, but for guidance in the great crises of human existence. This is the stone he offers to those whom he is bound by the strongest and tenderest ties to feed with bread.

If primary and secondary education are in this unsatisfactory state, what is to be said to the universities? This is an awful subject, and one I almost fear to touch with my unhallowed hands; but I can tell you what those say who have authority to speak.

The Rector of Lincoln College, in his lately published valuable " Suggestions for Academical Organisation with especial reference to Oxford," tells us (p. 127) :—

"The colleges were, in their origin, endow

ments, not for the elements of a general liberal
education, but for the prolonged stúdy of special
and professional faculties by men of riper age.
The universities embraced both these objects.
The colleges, while they incidentally aided in
elementary education, were specially devoted to
the highest learning. . . . .

"This was the theory of the middle-age university
and the design of collegiate foundations in their
origin. Time and circumstances have brought
about a total change. The colleges no longer
promote the researches of science, or direct professional study. Here and there college walls
may shelter an occasional student, but not in
larger proportions than may be found in private
life. Elementary teaching of youths under twenty
is now the only function performed by the university, and almost the only object of college endowments. Colleges were homes for the life-study of
the highest and most abstruse parts of knowledge.
They have become boarding schools in which the
elements of the learned languages are taught to
youths."

If Mr. Pattison's high position, and his obvious
love and respect for his university, be insufficient
to convince the outside world that language so
severe is yet no more than just, the authority of
the Commissioners who reported on the University
of Oxford in 1850 is open to no challenge. Yet
they write :—

"It is generally acknowledged that both Oxford and the country at large suffer greatly from the absence of a body of learned men devoting their lives to the cultivation of science, and to the direction of academical education.

"The fact that so few books of profound research emanate from the University of Oxford, materially impairs its character as a seat of learning, and consequently its hold on the respect of the nation."

Cambridge can claim no exemption from the reproaches addressed to Oxford. And thus there seems no escape from the admission that what we fondly call our great seats of learning are simply "boarding schools" for bigger boys; that learned men are not more numerous in them than out of them ; that the advancement of knowledge is not the object of fellows of colleges; that, in the philosophic calm and meditative stillness of their greenswarded courts, philosophy does not thrive, and meditation bears few fruits.

It is my great good fortune to reckon amongst my friends resident members of both universities, who are men of learning and research, zealous cultivators of science, keeping before their minds a noble ideal of a university, and doing their best to make that ideal a reality; and, to me, they would necessarily typify the universities, did not the authoritative statements I have quoted compel me to believe that they are exceptional, and not representative men. Indeed, upon calm consideration, several circumstances lead me to think that the Rector of Lincoln College and the Commissioners cannot be far wrong.

I believe there can be no doubt that the foreigner who should wish to become acquainted with the scientific, or the literary, activity of modern England, would simply lose his time and his pains if he visited ·our universities with that object.

And, as for works of profound research on any subject, and, above all, in that classical lore for which the universities profess to sacrifice almost everything else, why, a third-rate, poverty-stricken German university turns out more produce of that kind in one year, than our vast and wealthy foundations elaborate in ten.

Ask the man who is investigating any question, profoundly and thoroughly—be it historical, philosophical, philological, physical, literary, or theological; who is trying to make himself master of any abstract subject (except, perhaps, political economy and geology, both of which are intensely Anglican sciences), whether he is not compelled to read half a dozen times as many German as English books? And whether, of these English books, more than one in ten is the work of a fellow of a college, or a professor of an English university?

Is this from any lack of power in the English

as compared with the German mind? The
countrymen of Grote and of Mill, of Faraday, of
Robert Brown, of Lyell, and of Darwin, to go no
further back than the contemporaries of men of
middle age, can afford to smile at such a suggestion.
England can show now, as she has been able to
show in every generation since civilisation spread
over the West, individual men who hold their
own against the world, and keep alive the old
tradition of her intellectual eminence.

But, in the majority of cases, these men are
what they are in virtue of their native intellectual
force, and of a strength of character which will
not recognise impediments. They are not trained
in the courts of the Temple of Science, but storm
the walls of that edifice in all sorts of irregular
ways, and with much loss of time and power, in
order to obtain their legitimate positions.

Our universities not only do not encourage such
men; do not offer them positions, in which it
should be their highest duty to do, thoroughly,
that which they are most capable of doing; but,
as far as possible, university training shuts out of
the minds of those among them, who are subjected
to it, the prospect that there is anything in the
world for which they are specially fitted. Imagine
the success of the attempt to still the intellectual
hunger of any of the men I have mentioned, by
putting before him, as the object of existence, the
successful mimicry of the measure of a Greek

song, or the roll of Ciceronian prose! Imagine
how much success would be likely to attend the
attempt to persuade such men that the education
which leads to perfection in such elegances is
alone to be called culture; while the facts of
history, the process of thought, the conditions of
moral and social existence, and the laws of
physical nature are left to be dealt with as they
may by outside barbarians!

It is not thus that the German universities,
from being beneath notice a century ago, have
become what they are now—the most intensely
cultivated and the most productive intellectual
corporations the world has ever seen.

The student who repairs to them sees in the
list of classes and of professors a fair picture of
the world of knowledge. Whatever he needs to
know there is some one ready to teach him, some
one competent to discipline him in the way of
learning; whatever his special bent, let him but
be able and diligent, and in due time he shall
find distinction and a career. Among his professors, he sees men whose names are known and
revered throughout the civilised world; and their
living example infects him with a noble ambition,
and a love for the spirit of work.

The Germans dominate the intellectual world
by virtue of the same simple secret as that which
made Napoleon the master of old Europe. They
have declared *la carriere ouverte aux talents*, and

every Bursch marches with a professor's gown in
his knapsack. Let him become a great scholar,
or man of science, and ministers will compete for
his services. In Germany, they do not leave the
chance of his holding the office he would render
illustrious to the tender mercies of a hot canvass,
and the final wisdom of a mob of country
parsons.

In short, in Germany, the universities are
exactly what the Rector of Lincoln and the
Commissioners tell us the English universities are
not; that is to say, corporations "of learned men
devoting their lives to the cultivation of science,
and the direction of academical education."
They are not "boarding schools for youths," nor
clerical seminaries; but institutions for the higher
culture of men, in which the theological faculty is
of no more importance, or prominence, than the
rest; and which are truly "universities," since
they strive to represent and embody the totality
of human knowledge, and to find room for all
forms of intellectual activity.

May zealous and clear-headed reformers like
Mr. Pattison succeed in their noble endeavours to
shape our universities towards some such ideal as
this, without losing what is valuable and distinctive in their social tone! But until they have
succeeded, a liberal education will be no more
obtainable in our Oxford and Cambridge Universities than in our public schools.

If I am justified in my conception of the ideal of a liberal education ; and if what I have said about the existing educational institutions of the country is also true, it is clear that the two have no sort of relation to one another; that the best of our schools and the most complete of our university trainings give but a narrow, one-sided, and essentially illiberal education—while the worst give what is really next to no education at all.   The South London Working-Men's College could not copy any of these institutions if it would; I am bold enough to express the conviction that it ought not if it could.

For what is wanted is the reality and not the mere name of a liberal education ; and this College must steadily set before itself the ambition to be able to give that education sooner or later.   At present we are but beginning, sharpening our educational tools, as it were, and, except a modicum of physical science, we are not able to offer much more than is to be found in an ordinary school.

Moral and social science—one of the greatest and most fruitful of our future classes, I hope—at present lacks only one thing in our programme, and that is a teacher.   A considerable want, no doubt ; but it must be recollected that it is much better to want a teacher than to want the desire to learn.

Further, we need what, for want of a better

name, I must call Physical Geography. What I mean is that which the Germans call "*Erdkunde.*" It is a description of the earth, of its place and relation to other bodies; of its general structure, and of its great features—winds, tides, mountains, plains: of the chief forms of the vegetable and animal worlds, of the varieties of man. It is the peg upon which the greatest quantity of useful and entertaining scientific information can be suspended.

Literature is not upon the College programme; but I hope some day to see it there. For literature is the greatest of all sources of refined pleasure, and one of the great uses of a liberal education is to enable us to enjoy that pleasure. There is scope enough for the purposes of liberal education in the study of the rich treasures of our own language alone. All that is needed is direction, and the cultivation of a refined taste by attention to sound criticism. But there is no reason why French and German should not be mastered sufficiently to read what is worth reading in those languages with pleasure and with profit.

And finally, by and by, we must have History, treated not as a succession of battles and dynasties; not as a series of biographies; not as evidence that Providence has always been on the side of either Whigs or Tories; but as the development of man in times past, and in other conditions than our own.

tion, and have even thrown out timid and faltering
suggestions as to what should be done, while at
the opposite pole of society, committees of working
men have expressed their conviction that scientific
training is the one thing needful for their advancement, whether as men, or as workmen.   Only the
other day, it was my duty to take part in the
reception of a deputation of London working men,
who desired to learn from Sir Roderick Murchison,
the Director of the Royal School of Mines, whether
the organisation of the Institution in Jermyn Street
could be made available for the supply of that
scientific instruction the need of which could not
have been apprehended, or stated, more clearly than
it was by them.

The heads of colleges in our great universities
(who have not the reputation of being the most
mobile of persons) have, in several cases, thought
it well that, out of the great number of honours
and rewards at their disposal, a few should hereafter be given to the cultivators of the physical
sciences.   Nay, I hear that some colleges have even
gone so far as to appoint one, or, maybe, two special
tutors for the purpose of putting the facts and
principles of physical science before the undergraduate mind.   And I say it with gratitude and
great respect for those eminent persons, that the
head masters of our public schools, Eton, Harrow
Winchester, have addressed themselves to the
problem of introducing instruction in physical

science among the studies of those great educational bodies, with much honesty of purpose and enlightenment of understanding; and I live in hope that, before long, important changes in this direction will be carried into effect in those strongholds of ancient prescription. In fact, such changes have already been made, and physical science, even now, constitutes a recognised element of the school curriculum in Harrow and Rugby, whilst I understand that ample preparations for such studies are being made at Eton and elsewhere.

Looking at these facts, I might perhaps spare myself the trouble of giving any reasons for the introduction of physical science into elementary education; yet I cannot but think that it may be well if I place before you some considerations which, perhaps, have hardly received full attention.

At other times, and in other places, I have endeavoured to state the higher and more abstract arguments, by which the study of physical science may be shown to be indispensable to the complete training of the human mind; but I do not wish it to be supposed that, because I happen to be devoted to more or less abstract and "unpractical" pursuits, I am insensible to the weight which ought to be attached to that which has been said to be the English conception of Paradise—namely, "getting on." I look upon it, that "getting on" is a very important matter indeed. I do not mean merely for the sake of the coarse and tangible results of success, but because humanity is so constituted that a vast number of us would never be impelled to those stretches of exertion which make us wiser and more capable men, if it were not for the absolute necessity of putting on our faculties all the strain they will bear, for the purpose of ' getting on " in the most practical sense.

Now the value of a knowledge of physical science as a means of getting on is indubitable. There are hardly any of our trades, except the merely huckstering ones, in which some knowledge of science may not be directly profitable to the pursuer of that occupation. As industry attains higher stages of its development, as its processes become more complicated and refined, and competition more keen, the sciences are dragged in, one by one, to take their share in the fray; and he who can best avail himself of their help is the man who will come out uppermost in that struggle for existence, which goes on as fiercely beneath the smooth surface of modern society, as among the wild inhabitants of the woods.

But in addition to the bearing of science on ordinary practical life, let me direct your attention to its immense influence on several of the professions. I ask any one who has adopted the calling of an engineer, how much time he lost when he left school, because he had to devote himself to pursuits which were absolutely novel and strange,

and of which he had not obtained the remotest
conception from his instructors ?   He had to
familiarise himself with ideas of the course and
powers of Nature, to which his attention had never
been directed during his school-life, and to learn,
for the first time, that a world of facts lies outside
and beyond the world of words.   I appeal to those
who know what engineering is, to say how far I am
right in respect to that profession ; but with regard
to another, of no less importance, I shall venture
to speak of my own knowledge.   There is no one
of us who may not at any moment be thrown
bound hand and foot by physical incapacity, into
the hands of a medical practitioner.   The chances
of life and death for all and each of us may, at any
moment, depend on the skill with which that practitioner is able to make out what is wrong in our
bodily frames, and on his ability to apply the proper
remedy to the defect.

The necessities of modern life are such, and the
class from which the medical profession is chiefly
recruited is so situated, that few medical men can
hope to spend more than three or four, or it may
be five, years in the pursuit of those studies which
are immediately germane to physic.   How is that all
too brief period spent at present ?   I speak as an
old examiner, having served some eleven or twelve
years in that capacity in the University of London,
and therefore having a practical acquaintance with
the subject; but I might fortify myself by the

authority of the President of the College of
Surgeons, Mr. Quain, whom I heard the other day
in an admirable address (the Hunterian Oration)
deal fully and wisely with this very topic.[1]

A young man commencing the study of medicine
is at once required to endeavour to make an acquaintance with a number of sciences, such as
Physics, as Chemistry, as Botany, as Physiology,
which are absolutely and entirely strange to him,
however excellent his so-called education at school
may have been. Not only is he devoid of all
apprehension of scientific conceptions, not only does
he fail to attach any meaning to the words " mat-

---

[1] Mr. Quain's words (*Medical Times and Gazette*, February
20) are :—"A few words as to our special Medical course of
instruction and the influence upon it of such changes in the
elementary schools as I have mentioned. The student now
enters at once upon several sciences—physics, chemistry,
anatomy, physiology, botany, pharmacy, therapeutics—all these,
the facts and the language and the laws of each, to be mastered
in eighteen months   Up to the beginning of the Medical course
many have learned little.   We cannot claim anything better
than the Examiner of the University of London and the Cambridge Lecturer have reported for their Universities. Supposing
that at school young people had acquired some exact elementary
knowledge in physics, chemistry, and a branch of natural
history—say botany—with the physiology connected with it,
they would then have gained necessary knowledge, with some
practice in inductive reasoning. The whole studies are processes of observation and induction—the best discipline of the
mind for the purposes of life—for our purposes not less than
any. 'By such study (says Dr. Whewell) of one or more
departments of inductive science the mind may escape from the
thraldom of mere words.' By that plan the burden of the early
Medical course would be much lightened, and more time devoted
to practical studies, including Sir Thomas Watson's 'final and
supreme stage' of the knowledge of Medicine."

ter," "force," or "law" in their scientific senses, but, worse still, he has no notion of what it is to come into contact with Nature, or to lay his mind alongside of a physical fact, and try to conquer it, in the way our great naval hero told his captains to master their enemies. His whole mind has been given to books, and I am hardly exaggerating if I say that they are more real to him than Nature. He imagines that all knowledge can be got out of books, and rests upon the authority of some master or other; nor does he entertain any misgiving that the method of learning which led to proficiency in the rules of grammar will suffice to lead him to a mastery of the laws of Nature. The youngster, thus unprepared for serious study, is turned loose among his medical studies, with the result, in nine cases out of ten, that the first year of his curriculum is spent in learning how to learn. Indeed, he is lucky if, at the end of the first year, by the exertions of his teachers and his own industry, he has acquired even that art of arts. After which there remain not more than three, or perhaps four, years for the profitable study of such vast sciences as Anatomy, Physiology, Therapeutics, Medicine, Surgery, Obstetrics, and the like, upon his knowledge or ignorance of which it depends whether the practitioner shall diminish, or increase, the bills of mortality. Now what is it but the preposterous condition of ordinary school education which prevents a young man of seventeen, destined for the often that contact is to be described as collision, or violent friction ; and how great the heat, how little the light, which commonly results from it.

In the interests of fair play, to say nothing of those of mankind, I ask, Why do not the clergy as a body acquire, as a part of their preliminary education, some such tincture of physical science as will put them in a position to understand the difficulties in the way of accepting their theories, which are forced upon the mind of every thoughtful and intelligent man, who has taken the trouble to instruct himself in the elements of natural knowledge ?

Some time ago I attended a large meeting of the clergy, for the purpose of delivering an address which I had been invited to give. I spoke of some of the most elementary facts in physical science, and of the manner in which they directly contradict certain of the ordinary teachings of the clergy. The result was, that, after I had finished, one section of the assembled ecclesiastics attacked me with all the intemperance of pious zeal, for stating facts and conclusions which no competent judge doubts ; while, after the first speakers had subsided, amidst the cheers of the great majority of their colleagues, the more rational minority rose to tell me that I had taken wholly superfluous pains, that they already knew all about what I had told them, and perfectly agreed with me. A hard-headed friend of mine, who was present, put the not unnatural question, "Then why don't you say so in your pulpits?" to which inquiry I heard no reply.

In fact the clergy are at present divisible into three sections: an immense body who are ignorant and speak out; a small proportion who know and are silent; and a minute minority who know and speak according to their knowledge. By the clergy, I mean especially the Protestant clergy. Our great antagonist—I speak as a man of science—the Roman Catholic Church, the one great spiritual organisation which is able to resist, and must, as a matter of life and death, resist, the progress of science and modern civilisation, manages her affairs much better.

It was my fortune some time ago to pay a visit to one of the most important of the institutions in which the clergy of the Roman Catholic Church in these islands are trained; and it seemed to me that the difference between these men and the comfortable champions of Anglicanism and of Dissent, was comparable to the difference between our gallant Volunteers and the trained veterans of Napoleon's Old Guard.

The Catholic priest is trained to know his business, and do it effectually. The professors of the college in question, learned, zealous, and determined men, permitted me to speak frankly with them. We talked like outposts of opposed armies during a truce—as friendly enemies; and when I ventured to point out the difficulties their students would have to encounter from scientific thought, they replied : " Our Church has lasted many ages, and has passed safely through many storms. The present is but a new gust of the old tempest, and we do not turn out our young men less fitted to weather it, than they have been, in former times, to cope with the difficulties of those times. The heresies of the day are explained to them by their professors of philosophy and science, and they are taught how those heresies are to be met."

I heartily respect an organisation which faces its enemies in this way ; and I wish that all ecclesiastical organisations were in as effective a condition. I think it would be better, not only for them, but for us. The army of liberal thought is, at present, in very loose order ; and many a spirited free-thinker makes use of his freedom mainly to vent nonsense. We should be the better for a vigorous and watchful enemy to hammer us into cohesion and discipline ; and I, for one, lament that the bench of Bishops cannot show a man of the calibre of Butler of the " Analogy," who, if he were alive, would make short work of much of the current *a priori* " infidelity."

I hope you will consider that the arguments I have now stated, even if there were no better ones, constitute a sufficient apology for urging the introduction of science into schools.

The next question to which I have to address myself is, What sciences ought to be thus taught? And this is one of the most important of questions, because my side (I am afraid I am a terribly candid friend) sometimes spoils its cause by going in for too much. There are other forms of culture beside physical science; and I should be profoundly sorry to see the fact forgotten, or even to observe a tendency to starve, or cripple, literary, or æsthetic, culture for the sake of science. Such a narrow view of the nature of education has nothing to do with my firm conviction that a complete and thorough scientific culture ought to be introduced into all schools. By this, however, I do not mean that every schoolboy should be taught everything in science. That would be a very absurd thing to conceive, and a very mischievous thing to attempt. What I mean is, that no boy nor girl should leave school without possessing a grasp of the general character of science, and without having been disciplined, more or less, in the methods of all sciences; so that, when turned into the world to make their own way, they shall be prepared to face scientific problems, not by knowing at once the conditions of every problem, or by being able at once to solve it; but by being familiar with the general current of scientific thought, and by being able to apply the methods of science in the proper way, when they have acquainted themselves with the conditions of the special problem.

That is what I understand by scientific education. To furnish a boy with such an education,
it is by no means necessary that he should devote
his whole school existence to physical science : in
fact, no one would lament so one-sided a proceeding more than I. Nay more, it is not necessary
for him to give up more than a moderate share of
his time to such studies, if they be properly
selected and arranged, and if he be trained in
them in a fitting manner.

I conceive the proper course to be somewhat as
follows. To begin with, let every child be
instructed in those general views of the phænomena of Nature for which we have no exact
English name. The nearest approximation to a
name for what I mean, which we possess, is
"physical geography." The Germans have a
better, "Erdkunde" ("earth knowledge" or
"geology" in its etymological sense), that is to
say, a general knowledge of the earth, and what
is on it, in it, and about it. If any one who has
had experience of the ways of young children will
call to mind their questions, he will find that so
far as they can be put into any scientific category,
they come under this head of "Erdkunde." The
child asks, "What is the moon, and why does it
shine?" "What is this water, and where does it
run?" "What is the wind?" "What makes
this waves in the sea?" "Where does this animal
live, and what is the use of that plant?" And

if not snubbed and stunted by being told not to
ask foolish questions, there is no limit to the
intellectual craving of a young child ; nor any
bounds to the slow, but solid, accretion of knowledge and development of the thinking faculty in
this way. To all such questions, answers which
are necessarily incomplete, though true as far as
they go, may be given by any teacher whose ideas
represent real knowledge and not mere book
learning; and a panoramic view of Nature,
accompanied by a strong infusion of the scientific
habit of mind, may thus be placed within the
reach of every child of nine or ten.

After this preliminary opening of the eyes to
the great spectacle of the daily progress of
Nature, as the reasoning faculties of the child
grow, and he becomes familiar with the use of the
tools of knowledge—reading, writing, and elementary mathematics—he should pass on to
what is, in the more strict sense, physical science.
Now there are two kinds of physical science : the
one regards form and the relation of forms to
one another, the other deals with causes and
effects. In many of what we term sciences, these
two kinds are mixed up together; but systematic
botany is a pure example of the former kind, and
physics of the latter kind, of science. Every
educational advantage which training in physical
science can give is obtainable from the proper
study of these two ; and I should be contented,

for the present, if they, added to our " Erdkunde,"
furnished the whole of the scientific curriculum
of school. Indeed, I conceive it would be one of
the greatest boons which could be conferred upon
England, if henceforward every child in the
country were instructed in the general knowledge
of the things about it, in the elements of physics,
and of botany. But I should be still better
pleased if there could be added somewhat of
chemistry, and an elementary acquaintance with
human physiology.

So far as school education is concerned, I want
to go no further just now ; and I believe that
such instruction would make an excellent introduction to that preparatory scientific training which,
as I have indicated, is so essential for the successful pursuit of our most important professions.
But this modicum of instruction must be so given
as to ensure real knowledge and practical discipline.   If scientific education is to be dealt with
as mere bookwork, it will be better not to
attempt it, but to stick to the Latin Grammar
which makes no pretence to be anything but
bookwork.

If the great benefits of scientific training are
sought, it is essential that such training should be
real : that is to say, that the mind of the scholar
should be brought into direct relation with fact,
that he should not merely be told a thing, but
made to see by the use of his own intellect and

ability that the thing is so and no otherwise.
The great peculiarity of scientific training, that in
virtue of which it cannot be replaced by any
other discipline whatsoever, is this bringing of the
mind directly into contact with fact, and practising
the intellect in the completest form of induction ;
that is to say, in drawing conclusions from particular facts made known by immediate observation
of Nature.

The other studies which enter into ordinary
education do not discipline the mind in this way.
Mathematical training is almost purely deductive.
The mathematician starts with a few simple propositions, the proof of which is so obvious that they
are called self-evident, and the rest of his work
consists of subtle deductions from them.   The
teaching of languages, at any rate as ordinarily
practised, is of the same general nature,—authority
and tradition furnish the data, and the mental
operations of the scholar are deductive.

Again : if history be the subject of study, the
facts are still taken upon the evidence of tradition
and authority.   You cannot make a boy see the
battle of Thermopylæ for himself, or know, of his
own knowledge, that Cromwell once ruled England.
There is no getting into direct contact with natural
fact by this road ; there is no dispensing with
authority, but rather a resting upon it.

In all these respects, science differs from other
educational discipline, and prepares the scholar for

common life. What have we to do in every-day
life ? Most of the business which demands our
attention is matter of fact, which needs, in the first
place, to be accurately observed or apprehended;
in the second, to be interpreted by inductive and
deductive reasonings, which are altogether similar
in their nature to those employed in science. In
the one case, as in the other, whatever is taken for
granted is so taken at one's own peril; fact and
reason are the ultimate arbiters, and patience and
honesty are the great helpers out of difficulty.

But if scientific training is to yield its most
eminent results, it must. I repeat, be made practical.
That is to say, in explaining to a child the general
phænomena of Nature, you must, as far as possible,
give reality to your teaching by object-lessons; in
teaching him botany, he must handle the plants
and dissect the flowers for himself; in teaching
him physics and chemistry, you must not be
solicitous to fill him with information, but you
must be careful that what he learns he knows of
his own knowledge. Don't be satisfied with telling
him that a magnet attracts iron. Let him see
that it does; let him feel the pull of the one upon
the other for himself. And, especially, tell him
that it is his duty to doubt until he is compelled,
by the absolute authority of Nature, to believe that
which is written in books. Pursue this discipline
carefully and conscientiously, and you may make
sure that, however scanty may be the measure of

all kinds of teaching, but most essential, I am disposed to think, when the scholars are very young.
This condition is, that the teacher should himself
really and practically know his subject. If he does,
he will be able to speak of it in the easy language,
and with the completeness of conviction, with which
he talks of any ordinary every-day matter. If he
does not, he will be afraid to wander beyond the
limits of the technical phraseology which he has
got up; and a dead dogmatism, which oppresses,
or raises opposition, will take the place of the lively
confidence, born of personal conviction, which
cheers and encourages the eminently sympathetic
mind of childhood.

I have already hinted that such scientific training as we seek for may be given without making
any extravagant claim upon the time now devoted
to education. We ask only for " a most favoured
nation " clause in our treaty with the schoolmaster;
we demand no more than that science shall have
as much time given to it as any other single subject—say four hours a week in each class of an
ordinary school.

For the present, I think men of science would
be well content with such an arrangement as this;
but speaking for myself, I do not pretend to
believe that such an arrangement can be, or will
be, permanent. In these times the educational
tree seems to me to have its roots in the air, its leaves
and flowers in the ground , and, I confess, I should

very much like to turn it upside down, so that its roots might be solidly embedded among the facts of Nature, and draw thence a sound nutriment for the foliage and fruit of literature and of art.   No educational system can have a claim to permanence, unless it recognises the truth that education has two great ends to which everything else must be subordinated.   The one of these is to increase knowledge; the other is to develop the love of right and the hatred of wrong.

With wisdom and uprightness a nation can make its way worthily, and beauty will follow in the footsteps of the two, even if she be not specially invited; while there is perhaps no sight in the whole world more saddening and revolting than is offered by men sunk in ignorance of everything but what other men have written; seemingly devoid of moral belief or guidance; but with the sense of beauty so keen, and the power of expression so cultivated, that their sensual caterwauling may be almost mistaken for the music of the spheres.

At present, education is almost entirely devoted to the cultivation of the power of expression, and of the sense of literary beauty.   The matter of having anything to say, beyond a hash of other people's opinions, or of possessing any criterion of beauty, so that we may distinguish between the Godlike and the devilish, is left aside as of no moment.   I think I do not err in saying that if science were made a foundation of education, instead of being, at most, stuck on as cornice to the edifice, this state of things could not exist.

In advocating the introduction of physical science as a leading element in education, I by no means refer only to the higher schools. On the contrary, I believe that such a change is even more imperatively called for in those primary schools, in which the children of the poor are expected to turn to the best account the little time they can devote to the acquisition of knowledge. A great step in this direction has already been made by the establishment of science-classes under the Department of Science and Art,—a measure which came into existence unnoticed, but which will, I believe, turn out to be of more importance to the welfare of the people than many political changes over which the noise of battle has rent the air.

Under the regulations to which I refer, a schoolmaster can set up a class in one or more branches of science; his pupils will be examined, and the State will pay him, at a certain rate, for all who succeed in passing. I have acted as an examiner under this system from the beginning of its establishment, and this year I expect to have not fewer than a couple of thousand sets of answers to questions in Physiology, mainly from young people of the artisan class, who have been taught in the schools which are now scattered all over great Britain and Ireland. Some of my colleagues, who have to deal with subjects such as

Geometry, for which the present teaching power
is better organised, I understand are likely to
have three or four times as many papers.   So far
as my own subjects are concerned, I can undertake to say that a great deal of the teaching, the
results of which are before me in these examinations, is very sound and good ; and I think it is
in the power of the examiners, not only to keep
up the present standard, but to cause an almost
unlimited improvement.   Now what does this
mean ?   It means that by holding out a very
moderate inducement, the masters of primary
schools in many parts of the country have been
led to convert them into little foci of scientific
instruction ; and that they and their pupils have
contrived to find, or to make, time enough to carry
out this object with a very considerable degree of
efficiency.   That efficiency will, I doubt not, be
very much increased as the system becomes known
and perfected, even with the very limited leisure
left to masters and teachers on week-days.   And
this leads me to ask, Why should scientific teaching
be limited to week-days ?

Ecclesiastically-minded persons are in the habit
of calling things they do not like by very hard
names, and I should not wonder if they brand
the proposition I am about to make as blasphemous,
and worse.   But, not minding this, I venture to
ask, Would there really be anything wrong in
using part of Sunday for the purpose of instructing

those who have no other leisure, in a knowledge of the phænomena of Nature, and of man's relation to Nature?

I should like to see a scientific Sunday-school in every parish, not for the purpose of superseding any existing means of teaching the people the things that are for their good, but side by side with them. I cannot but think that there is room for all of us to work in helping to bridge over the great abyss of ignorance which lies at our feet.

And if any of the ecclesiastical persons to whom I have referred, object that they find it derogatory to the honour of the God whom they worship, to awaken the minds of the young to the infinite wonder and majesty of the works which they proclaim His, and to teach them those laws which must needs be His laws, and therefore of all things needful for man to know—I can only recommend them to be let blood and put on low diet. There must be something very wrong going on in the instrument of logic if it turns out such conclusions from such premises.

VI

SCIENCE AND CULTURE

[1880]

SIX years ago, as some of my present hearers may remember, I had the privilege of addressing a large assemblage of the inhabitants of this city, who had gathered together to do honour to the memory of their famous townsman, Joseph Priestley ; [1] and, if any satisfaction attaches to posthumous glory, we may hope that the manes of the burnt-out philosopher were then finally appeased.

No man, however, who is endowed with a fair share of common sense, and not more than a fair share of vanity, will identify either contemporary or posthumous fame with the highest good; and Priestley's life leaves no doubt that he, at any rate, set a much higher value upon the advancement of knowledge, and the promotion of that

[1] See the first essay in this volume.

freedom of thought which is at once the cause
and the consequence of intellectual progress.

Hence I am disposed to think that, if Priestley
could be amongst us to-day, the occasion of our
meeting would afford him even greater pleasure
than the proceedings which celebrated the centenary of his chief discovery. The kindly heart
would be moved, the high sense of social duty
would be satisfied, by the spectacle of well-earned
wealth, neither squandered in tawdry luxury and
vainglorious show, nor scattered with the careless
charity which blesses neither him that gives nor
him that takes, but expended in the execution of
a well-considered plan for the aid of present and
future generations of those who are willing to help
themselves.

We shall all be of one mind thus far. But it
is needful to share Priestley's keen interest in
physical science ; and to have learned, as he had
learned, the value of scientific training in fields
of inquiry apparently far remote from physical
science ; in order to appreciate, as he would have
appreciated, the value of the noble gift which Sir
Josiah Mason has bestowed upon the inhabitants
of the Midland district.

For us children of the nineteenth century,
however, the establishment of a college under the
conditions of Sir Josiah Mason's Trust, has a
significance apart from any which it could have
possessed a hundred years ago It appears to be

an indication that we are reaching the crisis of the battle, or rather of the long series of battles, which have been fought over education in a campaign which began long before Priestley's time, and will probably not be finished just yet.

In the last century, the combatants were the champions of ancient literature on the one side, and those of modern literature on the other; but, some thirty years[1] ago, the contest became complicated by the appearance of a third army, ranged round the banner of Physical Science.

I am not aware that any one has authority to speak in the name of this new host. For it must be admitted to be somewhat of a guerilla force, composed largely of irregulars, each of whom fights pretty much for his own hand. But the impressions of a full private, who has seen a good deal of service in the ranks, respecting the present position of affairs and the conditions of a permanent peace, may not be devoid of interest; and I do not know that I could make a better use of the present opportunity than by laying them before you.

From the time that the first suggestion to introduce physical science into ordinary education was

---

[1] The advocacy of the introduction of physical science into general education by George Combe and others commenced a good deal earlier; but the movement had acquired hardly any practical force before the time to which I refer.

timidly whispered, until now, the advocates of
scientific education have met with opposition of
two kinds. On the one hand, they have been
pooh-poohed by the men of business who pride
themselves on being the representatives of practi-
cality; while, on the other hand, they have been
excommunicated by the classical scholars, in their
capacity of Levites in charge of the ark of culture
and monopolists of liberal education.

The practical men believed that the idol whom
they worship—rule of thumb—has been the source
of the past prosperity, and will suffice for the
future welfare of the arts and manufactures.
They were of opinion that science is speculative
rubbish; that theory and practice have nothing
to do with one another; and that the scientific
habit of mind is an impediment, rather than an
aid, in the conduct of ordinary affairs.

I have used the past tense in speaking of the
practical men—for although they were very
formidable thirty years ago, I am not sure that
the pure species has not been extirpated. In fact,
so far as mere argument goes, they have been
subjected to such a *feu d'enfer* that it is a miracle
if any have escaped. But I have remarked that
your typical practical man has an unexpected
resemblance to one of Milton's angels. His
spiritual wounds, such as are inflicted by logical
weapons, may be as deep as a well and as wide as
a church door, but beyond shedding a few drops

of ichor, celestial or otherwise, he is no whit the worse. So, if any of these opponents be left, I will not waste time in vain repetition of the demonstrative evidence of the practical value of science; but knowing that a parable will sometimes penetrate where syllogisms fail to effect an entrance, I will offer a story for their consideration.

Once upon a time, a boy, with nothing to depend upon but his own vigorous nature, was thrown into the thick of the struggle for existence in the midst of a great manufacturing population. He seems to have had a hard fight, inasmuch as, by the time he was thirty years of age, his total disposable funds amounted to twenty pounds. Nevertheless, middle life found him giving proof of his comprehension of the practical problems he had been roughly called upon to solve, by a career of remarkable prosperity.

Finally, having reached old age with its well-earned surroundings of "honour, troops of friends," the hero of my story bethought himself of those who were making a like start in life, and how he could stretch out a helping hand to them.

After long and anxious reflection this successful practical man of business could devise nothing better than to provide them with the means of obtaining "sound, extensive, and practical scientific knowledge." And he devoted a large part of his wealth and five years of incessant work to this end.

I need not point the moral of a tale which, as
the solid and spacious fabric of the Scientific
College assures us, is no fable, nor can anything
which I could say intensify the force of this
practical answer to practical objections.

We may take it for granted then, that, in the
opinion of those best qualified to judge, the
diffusion of thorough scientific education is an
absolutely essential condition of industrial progress; and that the College which has been
opened to-day will confer an inestimable boon
upon those whose livelihood is to be gained by
the practise of the arts and manufactures of the
district.

The only question worth discussion is, whether
the conditions, under which the work of the
College is to be carried out, are such as to give it
the best possible chance of achieving permanent
success.

Sir Josiah Mason, without doubt most wisely,
has left very large freedom of action to the
trustees, to whom he proposes ultimately to
commit the administration of the College, so that
they may be able to adjust its arrangements in
accordance with the changing conditions of the
future. But, with respect to three points, he has
laid most explicit injunctions upon both administrators and teachers.

Party politics are forbidden to enter into the

minds of either, so far as the work of the College is concerned; theology is as sternly banished from its precincts; and finally, it is especially declared that the College shall make no provision for "mere literary instruction and education."

It does not concern me at present to dwell upon the first two injunctions any longer than may be needful to express my full conviction of their wisdom. But the third prohibition brings us face to face with those other opponents of scientific education, who are by no means in the moribund condition of the practical man, but alive, alert, and formidable.

It is not impossible that we shall hear this express exclusion of "literary instruction and education" from a College which, nevertheless, professes to give a high and efficient education, sharply criticised. Certainly the time was that the Levites of culture would have sounded their trumpets against its walls as against an educational Jericho.

How often have we not been told that the study of physical science is incompetent to confer culture; that it touches none of the higher problems of life; and, what is worse, that the continual devotion to scientific studies tends to generate a narrow and bigoted belief in the applicability of scientific methods to the search after truth of all kinds? How frequently one has reason to observe that no reply to a troublesome

argument tells so well as calling its author a
" mere scientific specialist." And, as I am afraid
it is not permissible to speak of this form of
opposition to scientific education in the past
tense; may we not expect to be told that this,
not only omission, but prohibition, of " mere
literary instruction and education" is a patent
example of scientific narrow-mindedness ?

I am not acquainted with Sir Josiah Mason's
reasons for the action which he has taken; but if,
as I apprehend is the case, he refers to the
ordinary classical course of our schools and
universities by the name of "mere literary instruction and education," I venture to offer
sundry reasons of my own in support of that
action.

For I hold very strongly by two convictions—
The first is, that neither the discipline nor the
subject-matter of classical education is of such
direct value to the student of physical science as
to justify the expenditure of valuable time upon
either; and the second is, that for the purpose of
attaining real culture, an exclusively scientific
education is at least as effectual as an exclusively
literary education.

I need hardly point out to you that these
opinions, especially the latter, are diametrically
opposed to those of the great majority of educated
Englishmen, influenced as they are by school and
university traditions. In their belief, culture is

obtainable only by a liberal education; and a liberal education is synonymous, not merely with education and instruction in literature, but in one particular form of literature, namely, that of Greek and Roman antiquity. They hold that the man who has learned Latin and Greek, however little, is educated; while he who is versed in other branches of knowledge, however deeply, is a more or less respectable specialist, not admissible into the cultured caste. The stamp of the educated man, the University degree, is not for him.

I am too well acquainted with the generous catholicity of spirit, the true sympathy with scientific thought, which pervades the writings of our chief apostle of culture to identify him with these opinions; and yet one may cull from one and another of those epistles to the Philistines, which so much delight all who do not answer to that name, sentences which lend them some support.

Mr. Arnold tells us that the meaning of culture is "to know the best that has been thought and said in the world." It is the criticism of life contained in literature. That criticism regards "Europe as being, for intellectual and spiritual purposes, one great confederation, bound to a joint action and working to a common result; and whose members have, for their common outfit, a knowledge of Greek, Roman, and Eastern antiquity, and of one another. Special, local, and temporary advantages being put out of account, that modern nation will in the intellectual and spiritual sphere make most progress, which most thoroughly carries out this programme. And what is that but saying that we too, all of us, as individuals, the more thoroughly we carry it out, shall make the more progress ? " [1]

We have here to deal with two distinct propositions. The first, that a criticism of life is the essence of culture; the second, that literature contains the materials which suffice for the construction of such a criticism.

I think that we must all assent to the first proposition. For culture certainly means something quite different from learning or technical skill. It implies the possession of an ideal, and the habit of critically estimating the value of things by comparison with a theoretic standard. Perfect culture should supply a complete theory of life, based upon a clear knowledge alike of its possibilities and of its limitations.

But we may agree to all this, and yet strongly dissent from the assumption that literature alone is competent to supply this knowledge. After having learnt all that Greek, Roman, and Eastern antiquity have thought and said, and all that modern literatures have to tell us, it is not self-evident that we have laid a sufficiently broad

[1] *Essays in Criticism*, p. 37.

and deep foundation for that criticism of life, which constitutes culture.

Indeed, to any one acquainted with the scope of physical science, it is not at all evident.   Considering progress only in the "intellectual and spiritual sphere," I find myself wholly unable to admit that either nations or individuals will really advance, if their common outfit draws nothing from the stores of physical science.   I should say that an army, without weapons of precision and with no particular base of operations, might more hopefully enter upon a campaign on the Rhine, than a man, devoid of a knowledge of what physical science has done in the last century, upon a criticism of life.

When a biologist meets with an anomaly, he in stinctively turns to the study of development to clear it up. The rationale of contradictory opinions may with equal confidence be sought in history.

It is, happily, no new thing that Englishmen should employ their wealth in building and endowing institutions for educational purposes. But, five or six hundred years ago, deeds of foundation expressed or implied conditions as nearly as possible contrary to those which have been thought expedient by Sir Josiah Mason. That is to say, physical science was practically ignored, while a certain literary training was en-joined as a means to the acquirement of knowledge which was essentially theological.

The reason of this singular contradiction between the actions of men alike animated by a strong and disinterested desire to promote the welfare of their fellows, is easily discovered.

At that time, in fact, if any one desired knowledge beyond such as could be obtained by his own observation, or by common conversation, his first necessity was to learn the Latin language, inasmuch as all the higher knowledge of the western world was contained in works written in that language. Hence, Latin grammar, with logic and rhetoric, studied through Latin, were the fundamentals of education. With respect to the substance of the knowledge imparted through this channel, the Jewish and Christian Scriptures, as interpreted and supplemented by the Romish Church, were held to contain a complete and infallibly true body of information.

Theological dicta were, to the thinkers of those days, that which the axioms and definitions of Euclid are to the geometers of these. The business of the philosophers of the middle ages was to deduce from the data furnished by the theologians, conclusions in accordance with ecclesiastical decrees. They were allowed the high privilege of showing, by logical process, how and why that which the Church said was true, must be true. And if their demonstrations fell short of or exceeded this limit, the Church was maternally ready to check their aberrations; if need were by the help of the secular arm.

Between the two, our ancestors were furnished with a compact and complete criticism of life. They were told how the world began and how it would end ; they learned that all material exist- ence was but a base and insignificant blot upon the fair face of the spiritual world, and that nature was, to all intents and purposes, the play-ground of the devil ; they learned that the earth is the centre of the visible universe, and that man is the cynosure of things terrestrial ; and more especially was it inculcated that the course of nature had no fixed order, but that it could be, and constantly was, altered by the agency of innumerable spiritual beings, good and bad, according as they were moved by the deeds and prayers of men. The sum and substance of the whole doctrine was to pro- duce the conviction that the only thing really worth knowing in this world was how to secure that place in a better which, under certain condi- tions, the Church promised.

Our ancestors had a living belief in this theory of life, and acted upon it in their dealings with education, as in all other matters. Culture meant saintliness—after the fashion of the saints of those days ; the education that led to it was, of necessity, theological ; and the way to theology lay through Latin.

That the study of nature—further than was re-

quisite for the satisfaction of everyday wants—
should have any bearing on human life was far
from the thoughts of men thus trained.   Indeed,
as nature had been cursed for man's sake, it was
an obvious conclusion that those who meddled with
nature were likely to come into pretty close contact
with Satan. And, if any born scientific investigator
followed his instincts, he might safely reckon upon
earning the reputation, and probably upon suffering the fate, of a sorcerer.

Had the western world been left to itself in
Chinese isolation, there is no saying how long this
state of things might have endured.   But, happily,
it was not left to itself.   Even earlier than the
thirteenth century, the development of Moorish
civilisation in Spain and the great movement of
the Crusades had introduced the leaven which,
from that day to this, has never ceased to work.
At first, through the intermediation of Arabic
translations, afterwards by the study of the originals, the western nations of Europe became
acquainted with the writings of the ancient philosophers and poets, and, in time, with the whole of
the vast literature of antiquity.

Whatever there was of high intellectual aspiration or dominant capacity in Italy, France,
Germany, and England, spent itself for centuries
in taking possession of the rich inheritance left
by the dead civilisations of Greece and Rome.
Marvellously aided by the invention of printing,

classical learning spread and flourished. Those who possessed it prided themselves on having attained the highest culture then within the reach of mankind.

And justly. For, saving Dante on his solitary pinnacle, there was no figure in modern literature at the time of the Renascence to compare with the men of antiquity; there was no art to compete with their sculpture; there was no physical science but that which Greece had created. Above all, there was no other example of perfect intellectual freedom—of the unhesitating acceptance of reason as the sole guide to truth and the supreme arbiter of conduct.

The new learning necessarily soon exerted a profound influence upon education. The language of the monks and schoolmen seemed little better than gibberish to scholars fresh from Virgil and Cicero, and the study of Latin was placed upon a new foundation. Moreover, Latin itself ceased to afford the sole key to knowledge. The student who sought the highest thought of antiquity, found only a second-hand reflection of it in Roman literature, and turned his face to the full light of the Greeks. And after a battle, not altogether dissimilar to that which is at present being fought over the teaching of physical science, the study of Greek was recognised as an essential element of all higher education.

Thus the Humanists, as they were called, won

the day; and the great reform which they effected was of incalculable service to mankind. But the Nemesis of all reformers is finality; and the reformers of education, like those of religion, fell into the profound, however common, error of mistaking the beginning for the end of the work of reformation.

The representatives of the Humanists, in the nineteenth century, take their stand upon classical education as the sole avenue to culture, as firmly as if we were still in the age of Renascence. Yet, surely, the present intellectual relations of the modern and the ancient worlds are profoundly different from those which obtained three centuries ago. Leaving aside the existence of a great and characteristically modern literature, of modern painting, and, especially, of modern music, there is one feature of the present state of the civilised world which separates it more widely from the Renascence, than the Renascence was separated from the middle ages.

This distinctive character of our own times lies in the vast and constantly increasing part which is played by natural knowledge. Not only is our daily life shaped by it, not only does the prosperity of millions of men depend upon it, but our whole theory of life has long been influenced, consciously or unconsciously, by the general conceptions of the universe, which have been forced upon us by physical science.

In fact, the most elementary acquaintance with the results of scientific investigation shows us that they offer a broad and striking contradiction to the opinion so implicitly credited and taught in the middle ages.

The notions of the beginning and the end of the world entertained by our forefathers are no longer credible. It is very certain that the earth is not the chief body in the material universe, and that the world is not subordinated to man's use. It is even more certain that nature is the expression of a definite order with which nothing interferes, and that the chief business of mankind is to learn that order and govern themselves accordingly. Moreover this scientific "criticism of life" presents itself to us with different credentials from any other. It appeals not to authority, nor to what anybody may have thought or said, but to nature. It admits that all our interpretations of natural fact are more or less imperfect and symbolic, and bids the learner seek for truth not among words but among things. It warns us that the assertion which outstrips evidence is not only a blunder but a crime.

The purely classical education advocated by the representatives of the Humanists in our day, gives no inkling of all this. A man may be a better scholar than Erasmus, and know no more of the chief causes of the present intellectual fermentation than Erasmus did. Scholarly and

pious persons, worthy of all respect, favour us
with allocutions upon the sadness of the antagonism of science to their mediæval way of thinking,
which betray an ignorance of the first principles
of scientific investigation, an incapacity for understanding what a man of science means by veracity,
and an unconsciousness of the weight of established scientific truths, which is almost comical.

There is no great force in the *tu quoque* argument, or else the advocates of scientific education
might fairly enough retort upon the modern
Humanists that they may be learned specialists,
but that they possess no such sound foundation
for a criticism of life as deserves the name of
culture. And, indeed, if we were disposed to be
cruel, we might urge that the Humanists have
brought this reproach upon themselves, not
because they are too full of the spirit of the
ancient Greek, but because they lack it.

The period of the Renascence is commonly
called that of the "Revival of Letters," as if the
influences then brought to bear upon the mind of
Western Europe had been wholly exhausted in
the field of literature.   I think it is very
commonly forgotten that the revival of science,
effected by the same agency, although less con-
spicuous, was not less momentous.

In fact, the few and scattered students of
nature of that day picked up the clue to her
secrets exactly as it fell from the hands of the

Greeks a thousand years before. The foundations of mathematics were so well laid by them, that our children learn their geometry from a book written for the schools of Alexandria two thousand years ago. Modern astronomy is the natural continuation and development of the work of Hipparchus and of Ptolemy; modern physics of that of Democritus and of Archimedes; it was long before modern biological science outgrew the knowledge bequeathed to us by Aristotle, by Theophrastus, and by Galen.

We cannot know all the best thoughts and sayings of the Greeks unless we know what they thought about natural phænomena. We cannot fully apprehend their criticism of life unless we understand the extent to which that criticism was affected by scientific conceptions. We falsely pretend to be the inheritors of their culture, unless we are penetrated, as the best minds among them were, with an unhesitating faith that the free employment of reason, in accordance with scientific method, is the sole method of reaching truth.

Thus I venture to think that the pretensions of our modern Humanists to the possession of the monopoly of culture and to the exclusive inheritance of the spirit of antiquity must be abated, if not abandoned. But I should be very sorry that anything I have said should be taken to imply a desire on my part to depreciate the value of classical education, as it might be and as it some-

times is. The native capacities of mankind vary
no less than their opportunities ; and while culture
is one, the road by which one man may best
reach it is widely different from that which is
most advantageous to another. Again, while
scientific education is yet inchoate and tentative,
classical education is thoroughly well organised
upon the practical experience of generations of
teachers. So that, given ample time for learning
and destination for ordinary life, or for a literary
career, I do not think that a young Englishman
in search of culture can do better than follow the
course usually marked out for him, supplementing
its deficiencies by his own efforts.

But for those who mean to make science their
serious occupation ; or who intend to follow the
profession of medicine ; or who have to enter early
upon the business of life; for all these, in my
opinion, classical education is a mistake ; and it is
for this reason that I am glad to see " mere
literary education and instruction " shut out from
the curriculum of Sir Josiah Mason's College,
seeing that its inclusion would probably lead to
the introduction of the ordinary smattering of
Latin and Greek.

Nevertheless, I am the last person to question
the importance of genuine literary education, or
to suppose that intellectual culture can be complete without it. An exclusively scientific training
will bring about a mental twist as surely as an

exclusively literary training   The value of the
cargo does not compensate for a ship's being out
of trim ; and I should be very sorry to think that
the Scientific College would turn out none but
lop-sided men.

There is no need, however, that such a catastrophe should happen.   Instruction in English,
French, and German is provided, and thus the
three greatest literatures of the modern world are
made accessible to the student.

French and German, and especially the latter
language, are absolutely indispensable to those
who desire full knowledge in any department of
science.   But even supposing that the knowledge
of these languages acquired is not more than
sufficient for purely scientific purposes, every
Englishman has, in his native tongue, an almost
perfect instrument of literary expression ; and, in
his own literature, models of every kind of literary
excellence.   If an Englishman cannot get literary
culture out of his Bible, his Shakespeare, his
Milton, neither, in my belief, will the profoundest
study of Homer and Sophocles, Virgil and Horace,
give it to him.

Thus, since the constitution of the College
makes sufficient provision for literary as well as
for scientific education, and since artistic instruction is also contemplated, it seems to me that a
fairly complete culture is offered to all who are
willing to take advantage of it.

But I am not sure that at this point the " practical " man, scotched but not slain, may ask what all this talk about culture has to do with an Institution, the object of which is defined to be " to promote the prosperity of the manufactures and the industry of the country." He may suggest that what is wanted for this end is not culture, nor even a purely scientific discipline, but simply a knowledge of applied science.

I often wish that this phrase, " applied science," had never been invented. For it suggests that there is a sort of scientific knowledge of direct practical use, which can be studied apart from another sort of scientific knowledge, which is of no practical utility, and which is termed " pure science." But there is no more complete fallacy than this. What people call applied science is nothing but the application of pure science to particular classes of problems. It consists of deductions from those general principles, established by reasoning and observation, which constitute pure science. No one can safely make these deductions until he has a firm grasp of the principles ; and he can obtain that grasp only by personal experience of the operations of observation and of reasoning on which they are founded.

Almost all the processes employed in the arts and manufactures fall within the range either of physics or of chemistry. In order to improve them, one must thoroughly understand them ; and no one has a chance of really understanding them, unless he has obtained that mastery of principles and that habit of dealing with facts, which is given by long continued and well-directed purely scientific training in the physical and the chemical laboratory. So that there really is no question as to the necessity of purely scientific discipline, even if the work of the College were limited by the narrowest interpretation of its stated aims.

And, as to the desirableness of a wider culture than that yielded by science alone, it is to be recollected that the improvement of manufacturing processes is only one of the conditions which contribute to the prosperity of industry. Industry is a means and not an end; and mankind work only to get something which they want. What that something is depends partly on their innate, and partly on their acquired, desires.

If the wealth resulting from prosperous industry is to be spent upon the gratification of unworthy desires, if the increasing perfection of manufacturing processes is to be accompanied by an increasing debasement of those who carry them on, I do not see the good of industry and prosperity.

Now it is perfectly true that men's views of what is desirable depend upon their characters; and that the innate proclivities to which we give that name are not touched by any amount of instruction. But it does not follow that even mere intellectual education may not, to an indefinite extent, modify the practical manifestation of the characters of men in their actions, by supplying them with motives unknown to the ignorant. A pleasure-loving character will have pleasure of some sort ; but, if you give him the choice, he may prefer pleasures which do not degrade him to those which do. And this choice is offered to every man, who possesses in literary or artistic culture a never-failing source of pleasures, which are neither withered by age, nor staled by custom, nor embittered in the recollection by the pangs of self-reproach.

If the Institution opened to-day fulfils the intention of its founder, the picked intelligences among all classes of the population of this district will pass through it.   No child born in Birmingham, henceforward, if he have the capacity to profit by the opportunities offered to him, first in the primary and other schools, and afterwards in the Scientific College, need fail to obtain, not merely the instruction, but the culture most appropriate to the conditions of his life.

Within these walls, the future employer and the future artisan may sojourn together for a while, and carry, through all their lives, the stamp of the influences then brought to bear upon them. Hence, it is not beside the mark to remind you, that the prosperity of industry depends not merely upon the improvement of manufacturing processes, not merely upon the ennobling of the individual char-

acter, but upon a third condition, namely, a clear
understanding of the conditions of social life, on
the part of both the capitalist and the operative,
and their agreement upon common principles of
social action.   They must learn that social phænomena are as much the expression of natural laws
as any others ; that no social arrangements can be
permanent unless they harmonise with the requirements of social statics and dynamics ; and that, in
the nature of things, there is an arbiter whose
decisions execute themselves.

But this knowledge is only to be obtained by the
application of the methods of investigation adopted
in physical researches to the investigation of the
phænomena of society.   Hence, I confess, I should
like to see one addition made to the excellent
scheme of education propounded for the College,
in the shape of provision for the teaching of
Sociology.   For though we are all agreed that
party politics are to have no place in the instruction of the College ; yet in this country, practically
governed as it is now by universal suffrage, every
man who does his duty must exercise political
functions.   And, if the evils which are inseparable
from the good of political liberty are to be checked,
if the perpetual oscillation of nations between
anarchy and despotism is to be replaced by the
steady march of self-restraining freedom ; it will
be because men will gradually bring themselves to
deal with political, as they now deal with scientific

questions; to be as ashamed of undue haste and partisan prejudice in the one case as in the other; and to believe that the machinery of society is at least as delicate as that of a spinning-jenny, and as little likely to be improved by the meddling of those who have not taken the trouble to master the principles of its action.

In conclusion, I am sure that I make myself the mouthpiece of all present in offering to the venerable founder of the Institution, which now commences its beneficent career, our congratulations on the completion of his work; and in expressing the conviction, that the remotest posterity will point to it as a crucial instance of the wisdom which natural piety leads all men to ascribe to their ancestors.

VII

ON SCIENCE AND ART IN RELATION TO EDUCATION

[1882]

WHEN a man is honoured by such a request as that which reached me from the authorities of your institution some time ago, I think the first thing that occurs to him is that which occurred to those who were bidden to the feast in the Gospel —to begin to make an excuse; and probably all the excuses suggested on that famous occasion crop up in his mind one after the other, including his "having married a wife," as reasons for not doing what he is asked to do. But, in my own case, and on this particular occasion, there were other difficulties of a sort peculiar to the time, and more or less personal to myself; because I felt that, if I came amongst you, I should be expected, and, indeed, morally compelled, to speak upon the subject of Scientific Education. And then there arose in my mind the recollection of a fact, which probably no one here but myself remembers; namely, that some fourteen years ago I was the guest of a citizen of yours, who bears the honoured name of Rathbone, at a very charming and pleasant dinner given by the Philomathic Society; and I there and then, and in this very city, made a speech upon the topic of Scientific Education. Under these circumstances, you see, one runs two dangers—the first, of repeating one's self, although I may fairly hope that everybody has forgotten the fact I have just now mentioned, except myself; and the second, and even greater difficulty, is the danger of saying something different from what one said before, because then, however forgotten your previous speech may be, somebody finds out its existence, and there goes on that process so hateful to members of Parliament, which may be denoted by the term "Hansardisation." Under these circumstances, I came to the conclusion that the best thing I could do was to take the bull by the horns, and to "Hansardise" myself,—to put before you, in the briefest possible way, the three or four propositions which I endeavoured to support on the occasion of the speech to which I have referred; and then to ask myself, supposing you were asking me, whether I had anything to retract, or to modify, in them, in virtue of the increased experience, and, let us charitably hope, the increased wisdom of an added fourteen years.

scientific culture ought to be introduced into all
schools."

I say I desire, in commenting upon these various
points, and judging them as fairly as I can by
the light of increased experience, to particularly
emphasise this last, because I am told, although I
assuredly do not know it of my own knowledge
—though I think if the fact were so I ought to
know it, being tolerably well acquainted with that
which goes on in the scientific world, and which
has gone on there for the last thirty years—that
there is a kind of sect, or horde, of scientific Goths
and Vandals, who think it would be proper and
desirable to sweep away all other forms of culture
and instruction, except those in physical science,
and to make them the universal and exclusive, or,
at any rate, the dominant training of the human
mind of the future generation. This is not my
view—I do not believe that it is anybody's view,
—but it is attributed to those who, like myself,
advocate scientific education. I therefore dwell
strongly upon the point, and I beg you to believe
that the words I have just now read were by no
means intended by me as a sop to the Cerberus of
culture. I have not been in the habit of offering
sops to any kind of Cerberus; but it was an
expression of profound conviction on my own part
—a conviction forced upon me not only by my
mental constitution, but by the lessons of what is

M 2

now becoming a somewhat long experience of
varied conditions of life.

I am not about to trouble you with my autobiography; the omens are hardly favourable, at
present, for work of that kind.   But I should like
if I may do so without appearing, what I earnestly
desire not to be, egotistical,—I should like to make
it clear to you, that such notions as these, which
are sometimes attributed to me, are, as I have said,
inconsistent with my mental constitution, and still
more inconsistent with the upshot of the teaching
of my experience.   For I can certainly claim for
myself that sort of mental temperament which can
say that nothing human comes amiss to it.   I
have never yet met with any branch of human
knowledge which I have found unattractive—
which it would not have been pleasant to me to
follow, so far as I could go ; and I have yet to
meet with any form of art in which it has
not been possible for me to take as acute a
pleasure as, I believe, it is possible for men to take.

And with respect to the circumstances of life, it
so happens that it has been my fate to know many
lands and many climates, and to be familiar, by
personal experience, with almost every form of
society, from the uncivilised savage of Papua
and Australia and the civilised savages of the
slums and dens of the poverty-stricken parts of
great cities, to those who perhaps, are occasionally

the somewhat over-civilised members of our upper ten thousand. And I have never found, in any of these conditions of life, a deficiency of something which was attractive. Savagery has its pleasures, I assure you, as well as civilisation, and I may even venture to confess—if you will not let a whisper of the matter get back to London, where I am known—I am even fain to confess, that sometimes in the din and throng of what is called "a brilliant reception" the vision crosses my mind of waking up from the soft plank which had afforded me satisfactory sleep during the hours of the night, in the bright dawn of a tropical morning, when my comrades were yet asleep, when every sound was hushed, except the little lap-lap of the ripples against the sides of the boat, and the distant twitter of the sea-bird on the reef. And when that vision crosses my mind, I am free to confess I desire to be back in the boat again. So that, if I share with those strange persons to whose asserted, but still hypothetical existence I have referred, the want of appreciation of forms of culture other than the pursuit of physical science, all I can say is, that it is, in spite of my constitution, and in spite of my experience, that such should be my fate.

But now let me turn to another point, or rather to two other points, with which I propose to occupy myself. How far does the experience of the last fourteen years justify the estimate which

I ventured to put forward of the value of scientific
culture, and of the share—the increasing share—
which it must take in ordinary education?
Happily, in respect to that matter, you need not
rely upon my testimony. In the last half-dozen numbers of the " Journal of Education," you will find
a series of very interesting and remarkable papers,
by gentlemen who are practically engaged in the
business of education in our great public and
other schools, telling us what is doing in these
schools, and what is their experience of the results
of scientific education there, so far as it has gone.
I am not going to trouble you with an abstract of
those papers, which are well worth your study in
their fulness and completeness, but I have copied
out one remarkable passage, because it seems to
me so entirely to bear out what I have formerly
ventured to say about the value of science, both as
to its subject-matter and as to the discipline which
the learning of science involves. It is from a
paper by Mr. Worthington—one of the masters at
Clifton, the reputation of which school you know
well, and at the head of which is an old friend of
mine, the Rev. Mr. Wilson—to whom much credit
is due for being one of the first, as I can say
from my own knowledge, to take up this question
and work it into practical shape. What Mr.
Worthington says is this :—

" It is not easy to exaggerate the importance of the information imparted by certain branches of science ; it modifies the

whole criticism of life made in maturer years. The study has
often, on a mass of boys, a certain influence which, I think, was
hardly anticipated, and to which a good deal of value must be
attached—an influence as much moral as intellectual, which is
shown in the increased and increasing respect for precision of
statement, and for that form of veracity which consists in the
acknowledgment of difficulties. It produces a real effect to find
that Nature cannot be imposed upon, and the attention given
to experimental lectures, at first superficial and curious only,
soon becomes minute, serious, and practical."

Ladies and gentlemen, I could not have chosen
better words to express—in fact, I have, in other
words, expressed the same conviction in former
days—what the influence of scientific teaching, if
properly carried out, must be.

But now comes the question of properly carrying
it out, because, when I hear the value of school
teaching in physical science disputed, my first im-
pulse is to ask the disputer, " What have you
known about it ? " and he generally tells me some
lamentable case of failure. Then I ask, " What are
the circumstances of the case, and how was the
teaching carried out ? " I remember, some few
years ago, hearing of the head master of a large
school, who had expressed great dissatisfaction
with the adoption of the teaching of physical
science—and that after experiment. But the experiment consisted in this—in asking one of the junior
masters in the school to get up science, in order to
teach it ; and the young gentleman went away for a
year and got up science and taught it. Well, I have

no doubt that the result was as disappointing as the head-master said it was, and I have no doubt that it ought to have been as disappointing, and far more disappointing too ; for, if this kind of instruction is to be of any good at all, if it is not to be less than no good, if it is to take the place of that which is already of some good, then there are several points which must be attended to.

And the first of these is the proper selection of topics, the second is practical teaching, the third is practical teachers, and the fourth is sufficiency of time. If these four points are not carefully attended to by anybody who undertakes the teaching of physical science in schools, my advice to him is, to let it alone. I will not dwell at any length upon the first point, because there is a general consensus of opinion as to the nature of the topics which should be chosen. The second point—practical teaching—is one of great importance, because it requires more capital to set it agoing, demands more time, and, last, but by no means least, it requires much more personal exertion and trouble on the part of those professing to teach, than is the case with other kinds of instruction.

When I accepted the invitation to be here this evening, your secretary was good enough to send me the addresses which have been given by distinguished persons who have previously occupied this chair. I don't know whether he had a malicious desire to alarm me ; but, however that

may be, I read the addresses, and derived the
greatest pleasure and profit from some of them,
and from none more than from the one given by
the great historian, Mr. Freeman, which delighted
me most of all; and, if I had not been ashamed of
plagiarising, and if I had not been sure of being
found out, I should have been glad to have copied
very much of what Mr. Freeman said, simply
putting in the word science for history. There
was one notable passage,—" The difference between good and bad teaching mainly consists in
this, whether the words used are really clothed
with a meaning or not." And Mr. Freeman gives
a remarkable example of this. He says, when a
little girl was asked where Turkey was, she
answered that it was in the yard with the other
fowls, and that showed she had a definite idea
connected with the word Turkey, and was, so far,
worthy of praise. I quite agree with that com-
mendation; but what a curious thing it is that
one should now find it necessary to urge that this
is the be-all and end-all of scientific instruction--
the *sine quâ non*, the absolutely necessary condition,
—and yet that it was insisted upon more than two
hundred years ago by one of the greatest men
science ever possessed in this country, William
Harvey. Harvey wrote, or at least published,
only two small books, one of which is the well-
known treatise on the circulation of the blood.
The other, the " Exercitationes de Generatione," is

less known, but not less remarkable. And not the least valuable part of it is the preface, in which there occurs this passage : "Those who, reading the words of authors, do not form sensible images of the things referred to, obtain no true ideas, but conceive false imaginations and inane phantasms." You see, William Harvey's words are just the same in substance as those of Mr. Freeman, only they happen to be rather more than two centuries older. So that what I am now saying has its application elsewhere than in science ; but assuredly in science the condition of knowing, of your own knowledge, things which you talk about, is absolutely imperative.

I remember, in my youth, there were detestable books which ought to have been burned by the hands of the common hangman, for they contained questions and answers to be learned by heart, of this sort, " What is a horse ? The horse is termed *Equus caballus* ; belongs to the class Mammalia ; order, Pachydermata ; family, Solidungula." Was any human being wiser for learning that magic formula ? Was he not more foolish, inasmuch as he was deluded into taking words for knowledge ? It is that kind of teaching that one wants to get rid of, and banished out of science. Make it as little as you like, but, unless that which is taught is based on actual observation and familiarity with facts, it is better left alone.

There are a great many people who imagine that

of it, it is needful you should familiarise yourself
with so much as you are called upon to teach—
soak yourself in it, so to speak—until you know it
as part of your daily life and daily knowledge, and
then you will be able to teach anybody. That is
what I mean by practical teachers, and, although
the deficiency of such teachers is being remedied
to a large extent, I think it is one which has long
existed, and which has existed from no fault of
those who undertook to teach, but because, until
the last score of years, it absolutely was not possible for any one in a great many branches of science,
whatever his desire might be, to get instruction
which would enable him to be a good teacher of elementary things. All that is being rapidly altered,
and I hope it will soon become a thing of the past.

The last point I have referred to is the question
of the sufficiency of time. And here comes the
rub. The teaching of science needs time, as any
other subject; but it needs more time proportion-
ally than other subjects, for the amount of work
obviously done, if the teaching is to be, as I have
said, practical. Work done in a laboratory involves
a good deal of expenditure of time without always
an obvious result, because we do not see anything
of that quiet process of soaking the facts into the
mind, which takes place through the organs of the
senses. On this ground there must be ample time
given to science teaching. What that amount
of time should be is a point which I need not

discuss now ; in fact, it is a point which cannot be settled until one has made up one's mind about various other questions.

All, then, that I have to ask for, on behalf of the scientific people, if I may venture to speak for more than myself, is that you should put scientific teaching into what statesmen call the condition of " the most favoured nation " ; that is to say, that it shall have as large a share of the time given to education as any other principal subject. You may say that that is a very vague statement, because the value of the allotment of time, under those circumstances, depends upon the number of principal subjects. It is x the time, and an unknown quantity of principal subjects dividing that, and science taking shares with the rest. That shows that we cannot deal with this question fully until we have made up our minds as to what the principal subjects of education ought to be.

I know quite well that launching myself into this discussion is a very dangerous operation ; that it is a very large subject, and one which is difficult to deal with, however much I may trespass upon your patience in the time allotted to me. But the discussion is so fundamental, it is so completely impossible to make up one's mind on these matters until one has settled the question, that I will even venture to make the experiment. A great lawyer-statesman and philosopher of a former age—I mean Francis Bacon—said that truth came out of error much more rapidly than it came out of confusion. There is a wonderful truth in that saying. Next to being right in this world, the best of all things is to be clearly and definitely wrong, because you will come out somewhere. If you go buzzing about between right and wrong, vibrating and fluctuating, you come out nowhere; but if you are absolutely and thoroughly and persistently wrong, you must, some of these days, have the extreme good fortune of knocking your head against a fact, and that sets you all straight again. So I will not trouble myself as to whether I may be right or wrong in what I am about to say, but at any rate I hope to be clear and definite; and then you will be able to judge for yourselves whether, in following out the train of thought I have to introduce, you knock your heads against facts or not.

I take it that the whole object of education is, in the first place, to train the faculties of the young in such a manner as to give their possessors the best chance of being happy and useful in their generation; and, in the second place, to furnish them with the most important portions of that immense capitalised experience of the human race which we call knowledge of various kinds. I am using the term knowledge in its widest possible sense; and the question is, what subjects to select by training and discipline, in which the object I have just defined may be best attained.

Now, it is a very remarkable fact—but it is
true of most things in this world—that there is
hardly anything one-sided, or of one nature ; and
it is not immediately obvious what of the things
that interest us may be regarded as pure science,
and what may be regarded as pure art. It may
be that there are some peculiarly constituted
persons who, before they have advanced far into
the depths of geometry, find artistic beauty about
it ; but, taking the generality of mankind, I think
it may be said that, when they begin to learn
mathematics, their whole souls are absorbed in
tracing the connection between the premises and
the conclusion, and that to them geometry is pure
science. So I think it may be said that mechanics
and osteology are pure science.   On the other
hand, melody in music is pure art.   You cannot
reason about it ; there is no proposition involved
in it. So, again, in the pictorial art, an arabesque,
or a " harmony in grey," touches none but the
æsthetic faculty.   But a great mathematician,
and even many persons who are not great mathematicians, will tell you that they derive immense
pleasure from geometrical reasonings. Everybody
knows mathematicians speak of solutions and
problems as " elegant," and they tell you that a
certain mass of mystic symbols is " beautiful,
quite lovely." Well, you do not see it. They do
see it, because the intellectual process, the process
of comprehending the reasons symbolised by these

figures and these signs, confers upon them a sort of pleasure, such as an artist has in visual symmetry. Take a science of which I may speak with more confidence, and which is the most attractive of those I am concerned with. It is what we call morphology, which consists in tracing out the unity in variety of the infinitely diversified structures of animals and plants. I cannot give you any example of a thorough æsthetic pleasure more intensely real than a pleasure of this kind—the pleasure which arises in one's mind when a whole mass of different structures run into one harmony as the expression of a central law. That is where the province of art overlays and embraces the province of intellect. And, if I may venture to express an opinion on such a subject, the great majority of forms of art are not in the sense what I just now defined them to be—pure art; but they derive much of their quality from simultaneous and even unconscious excitement of the intellect.

When I was a boy, I was very fond of music, and I am so now; and it so happened that I had the opportunity of hearing much good music. Among other things, I had abundant opportunities of hearing that great old master, Sebastian Bach. I remember perfectly well—though I knew nothing about music then, and, I may add, know nothing whatever about it now—the intense satisfaction and delight which I had in listening, by the hour together, to Bach's fugues. It is a pleasure which remains with me, I am glad to think; but, of late years, I have tried to find out the why and wherefore, and it has often occurred to me that the pleasure derived from musical compositions of this kind is essentially of the same nature as that which is derived from pursuits which are commonly regarded as purely intellectual. I mean, that the source of pleasure is exactly the same as in most of my problems in morphology—that you have the theme in one of the old master's works followed out in all its endless variations, always appearing and always reminding you of unity in variety. So in painting; what is called " truth to nature " is the intellectual element coming in, and truth to nature depends entirely upon the intellectual culture of the person to whom art is addressed. If you are in Australia, you may get credit for being a good artist—I mean among the natives— if you can draw a kangaroo after a fashion. But, among men of higher civilisation, the intellectual knowledge we possess brings its criticism into our appreciation of works of art, and we are obliged to satisfy it, as well as the mere sense of beauty in colour and in outline. And so, the higher the culture and information of those whom art addresses, the more exact and precise must be what we call its " truth to nature."

If we turn to literature, the same thing is true,

and you find works of literature which may be
said to be pure art. A little song of Shakespeare
or of Goethe is pure art; it is exquisitely beautiful,
although its intellectual content may be nothing.
A series of pictures is made to pass before your
mind by the meaning of words, and the effect is a
melody of ideas. Nevertheless, the great mass of
the literature we esteem is valued, not merely
because of having artistic form, but because of its
intellectual content; and the value is the higher
the more precise, distinct, and true is that intellectual content. And, if you will let me for a
moment speak of the very highest forms of
literature, do we not regard them as highest
simply because the more we know the truer they
seem, and the more competent we are to appreciate beauty the more beautiful they are ? No
man ever understands Shakespeare until he is old,
though the youngest may admire him, the reason
being that he satisfies the artistic instinct of the
youngest and harmonises with the ripest and
richest experience of the oldest.

I have said this much to draw your attention
to what, to my mind, lies at the root of all this
matter, and at the understanding of one another
by the men of science on the one hand, and the
men of literature, and history, and art, on the
other. It is not a question whether one order of
study or another should predominate. It is a
question of what topics of education you shall

select which will combine all the needful elements
in such due proportion as to give the greatest
amount of food, support, and encouragement
to those faculties which enable us to appreciate
truth, and to profit by those sources of innocent
happiness which are open to us, and, at the same
time, to avoid that which is bad, and coarse, and
ugly, and keep clear of the multitude of pitfalls
and dangers which beset those who break through
the natural or moral laws.

I address myself, in this spirit, to the consideration of the question of the value of purely literary
education. Is it good and sufficient, or is it
insufficient and bad ? Well, here I venture to
say that there are literary educations and literary
educations. If I am to understand by that term
the education that was current in the great
majority of middle-class schools, and upper schools
too, in this country when I was a boy, and which
consisted absolutely and almost entirely in keeping
boys for eight or ten years at learning the rules of
Latin and Greek grammar, construing certain
Latin and Greek authors, and possibly making
verses which, had they been English verses,
would have been condemned as abominable
doggerel,—if that is what you mean by liberal
education, then I say it is scandalously insufficient
and almost worthless. My reason for saying so
is not from the point of view of science at all, but
from the point of view of literature. I say the

thing professes to be literary education that is not a literary education at all. It was not literature at all that was taught, but science in a very bad form. It is quite obvious that grammar is science and not literature. The analysis of a text by the help of the rules of grammar is just as much a scientific operation as the analysis of a chemical compound by the help of the rules of chemical analysis. There is nothing that appeals to the æsthetic faculty in that operation; and I ask multitudes of men of my own age, who went through this process, whether they ever had a conception of art or literature until they obtained it for themselves after leaving school? Then you may say, "If that is so, if the education was scientific, why cannot you be satisfied with it?" I say, because although it is a scientific training, it is of the most inadequate and inappropriate kind. If there is any good at all in scientific education it is that men should be trained, as I said before, to know things for themselves at first hand, and that they should understand every step of the reason of that which they do.

I desire to speak with the utmost respect of that science—philology—of which grammar is a part and parcel; yet everybody knows that grammar, as it is usually learned at school, affords no scientific training. It is taught just as you would teach the rules of chess or draughts. On the other hand, if I am to understand by a literary

education the study of the literatures of either
ancient or modern nations—but especially those of
antiquity, and especially that of ancient Greece;
if this literature is studied, not merely from the
point of view of philological science, and its
practical application to the interpretation of texts,
but as an exemplification of and commentary
upon the principles of art; if you look upon the
literature of a people as a chapter in the development of the human mind, if you work out this in
a broad spirit, and with such collateral references
to morals and politics, and physical geography,
and the like as are needful to make you comprehend what the meaning of ancient literature and
civilisation is,—then, assuredly, it affords a
splendid and noble education. But I still think
it is susceptible of improvement, and that no man
will ever comprehend the real secret of the difference between the ancient world and our present
time, unless he has learned to see the difference
which the late development of physical science
has made between the thought of this day and the
thought of that, and he will never see that
difference, unless he has some practical insight
into some branches of physical science; and you
must remember that a literary education such as
that which I have just referred to, is out of the
reach of those whose school life is cut short at
sixteen or seventeen.

But, you will say, all this is fault-finding; let

us hear what you have in the way of positive
suggestion. Then I am bound to tell you that, if
I could make a clean sweep of everything—I am
very glad I cannot because I might, and
probably should, make mistakes,—but if I could
make a clean sweep of everything and start
afresh, I should, in the first place, secure that
training of the young in reading and writing, and
in the habit of attention and observation, both to
that which is told them, and that which they see,
which everybody agrees to. But in addition to
that, I should make it absolutely necessary for
everybody, for a longer or shorter period, to learn
to draw. Now, you may say, there are some
people who cannot draw, however much they may
be taught. I deny that *in toto*, because I never yet
met with anybody who could not learn to write.
Writing is a form of drawing; therefore if you
give the same attention and trouble to drawing
as you do to writing, depend upon it, there is
nobody who cannot be made to draw, more or less
well. Do not misapprehend me. I do not say
for one moment you would make an artistic
draughtsman. Artists are not made; they grow.
You may improve the natural faculty in that
direction, but you cannot make it; but you can
teach simple drawing, and you will find it an
implement of learning of extreme value. I do
not think its value can be exaggerated, because it
gives you the means of training the young in

beauty and of models of literary excellence which
exists in the world at the present time.   I have
said before, and I repeat it here, that if a man
cannot get literary culture of the highest kind out
of his Bible, and Chaucer, and Shakespeare, and
Milton, and Hobbes, and Bishop Berkeley, to
mention only a few of our illustrious writers—I
say, if he cannot get it out of those writers, he
cannot get it out of anything ; and I would
assuredly devote a very large portion of the time
of every English child to the careful study of the
models of English writing of such varied and
wonderful kind as we possess, and, what is still
more important and still more neglected, the habit
of using that language with precision, with force,
and with art.   I fancy we are almost the only
nation in the world who seem to think that composition comes by nature.   The French attend to
their own language, the Germans study theirs ; but
Englishmen do not seem to think it is worth their
while.   Nor would I fail to include, in the course
of study I am sketching, translations of all the
best works of antiquity, or of the modern world.
It is a very desirable thing to read Homer in
Greek ; but if you don't happen to know Greek,
the next best thing we can do is to read as good
a translation of it as we have recently been
furnished with in prose.   You won't get all you
would get from the original, but you may get a
great deal ; and to refuse to know this great deal

because you cannot get all, seems to be as sensible
as for a hungry man to refuse bread because he
cannot get partridge.   Finally, I would add instruction in either music or painting, or, if the
child should be so unhappy, as sometimes happens,
as to have no faculty for either of those, and no
possibility of doing anything in any artistic sense
with them, then I would see what could be done
with literature alone ; but I would provide, in the
fullest sense, for the development of the æsthetic
side of the mind.   In my judgment, those are all
the essentials of education for an English child.
With that outfit, such as it might be made in the
time given to education which is within the
reach of nine-tenths of the population—with that
outfit, an Englishman, within the limits of
English life, is fitted to go anywhere, to
occupy the highest positions, to fill the highest
offices   of   the   State,   and   to   become   distinguished in practical pursuits, in science, or in
art.   For, if he have the opportunity to learn all
those things, and have his mind disciplined in
the various directions the teaching of those topics
would have necessitated, then, assuredly, he will
be able to pick up, on his road through life, all the
rest of the intellectual baggage he wants.

If the educational time at our disposition were
sufficient, there are one or two things I would add
to those I have just now called the essentials ; and
perhaps you will be surprised to hear, though I

hope you will not, that I should add, not more
science, but one, or, if possible, two languages.
The knowledge of some other language than one's
own is, in fact, of singular intellectual value.
Many of the faults and mistakes of the ancient
philosophers are traceable to the fact that they
knew no language but their own, and were often
led into confusing the symbol with the thought
which it embodied.   I think it is Locke who says
that one-half of the mistakes of philosophers have
arisen from questions about words; and one of the
safest ways of delivering yourself from the bondage
of words is, to know how ideas look in words to
which you are not accustomed.   That is one reason
for the study of language; another reason is, that
it opens new fields in art and in science.   Another
is the practical value of such knowledge; and yet
another is this, that if your languages are properly
chosen, from the time of learning the additional
languages you will know your own language better
than ever you did.   So, I say, if the time given
to education permits, add Latin and German.
Latin, because it is the key to nearly one-half of
English and to all the Romance languages; and
German, because it is the key to almost all the
remainder of English, and helps you to understand
a race from whom most of us have sprung, and
who have a character and a literature of a fateful
force in the history of the world, such as probably
has been allotted to those of no other people,

VIII

UNIVERSITIES : ACTUAL AND IDEAL

[1874]

ELECTED by the suffrages of your four Nations Rector of the ancient University of which you are scholars, I take the earliest opportunity which has presented itself since my restoration to health, of delivering the Address which, by long custom, is expected of the holder of my office.

My first duty in opening that Address, is to offer you my most hearty thanks for the signal honour you have conferred upon me—an honour of which, as a man unconnected with you by personal or by national ties, devoid of political distinction, and a plebeian who stands by his order, I could not have dreamed. And it was the more surprising to me, as the five-and-twenty years which have passed over my head since I reached intellectual manhood, have been largely spent in no half-hearted advocacy of doctrines which have not yet found favour in the eyes of Academic respectability; so that, when the proposal to nominate me for your Rector came, I was almost as much astonished as was Hal o' the Wynd, "who fought for his own hand," by the Black Douglas's proffer of knighthood.   And I fear that my acceptance must be taken as evidence that, less wise than the Armourer of Perth, I have not yet done with soldiering.

In fact, if, for a moment, I imagined that your intention was simply, in the kindness of your hearts, to do me honour; and that the Rector of your University, like that of some other Universities was one of those happy beings who sit in glory for three years, with nothing to do for it save the making of a speech, a conversation with my distinguished predecessor soon dispelled the dream.   I found that, by the constitution of the University of Aberdeen, the incumbent of the Rectorate is, if not a power, at any rate a potential energy; and that, whatever may be his chances of success or failure, it is his duty to convert that potential energy into a living force, directed towards such ends as may seem to him conducive to the welfare of the corporation of which he is the theoretical head.

I need not tell you that your late Lord Rector took this view of his position, and acted upon it with the comprehensive, far-seeing insight into the actual condition and tendencies, not merely

of his own, but of other countries, which is his
honourable characteristic among statesmen.   I
have already done my best, and, as long as I hold
my office, I shall continue my endeavours, to follow
in the path which he trod ; to do what in me lies,
to bring this University nearer to the ideal—alas,
that I should be obliged to say ideal—of all
Universities ; which, as I conceive, should be places
in which thought is free from all fetters ; and in
which all sources of knowledge, and all aids to
learning, should be accessible to all comers, without distinction of creed or country, riches or
poverty.

Do not suppose, however, that I am sanguine
enough to expect much to come of any poor efforts
of mine.   If your annals take any notice of my
incumbency, I shall probably go down to posterity
as the Rector who was always beaten.   But if they
add, as I think they will, that my defeats became
victories in the hands of my successors, I shall be
well content.

The scenes are shifting in the great theatre of the
world.   The act which commenced with the Protestant Reformation is nearly played out, and a wider
and deeper change than that effected three centuries ago— a reformation, or rather a revolution of
thought, the extremes of which are represented by
the intellectual heirs of John of Leyden and of
Ignatius Loyola, rather than by those of Luther

and of Leo—is waiting to come on, nay, visible
behind the scenes to those who have good eyes
Men are beginning, once more, to awake to the
fact that matters of belief and of speculation are of
absolutely infinite practical importance; and are
drawing off from that sunny country " where it is
always afternoon"—the sleepy hollow of broad
indifferentism—to range themselves under their
natural banners.   Change is in the air.   It is
whirling feather-heads into all sorts of eccentric
orbits, and filling the steadiest with a sense of in-
security.   It insists on reopening all questions and
asking all institutions, however venerable, by what
right they exist, and whether they are, or are not,
in harmony with the real or supposed wants of
mankind.   And it is remarkable that these searching inquiries are not so much forced on institutions from without, as developed from within.
Consummate scholars question the value of learning; priests contemn dogma; and women turn
their backs upon man's ideal of perfect womanhood, and seek satisfaction in apocalyptic visions
of some, as yet, unrealised epicene reality.

If there be a type of stability in this world, one
would be inclined to look for it in the old Universities of England.   But it has been my business
of late to hear a good deal about what is going on
in these famous corporations; and I have been
filled with astonishment by the evidences of internal fermentation which they exhibit.   If Gibbon

could revisit the ancient seat of learning of which
he has written so cavalierly, assuredly he would
no longer speak of "the monks of Oxford sunk in
prejudice and port." There, as elsewhere, port
has gone out of fashion, and so has prejudice—at
least that particular fine, old, crusted sort of prejudice to which the great historian alludes.

Indeed, things are moving so fast in Oxford and
Cambridge, that, for my part, I rejoiced when the
Royal Commission, of which I am a member, had
finished and presented the Report which related
to these Universities; for we should have looked
like mere plagiarists, if, in consequence of a little
longer delay in issuing it, all the measures of
reform we proposed had been anticipated by the
spontaneous action of the Universities themselves.

A month ago I should have gone on to say that
one might speedily expect changes of another
kind in Oxford and Cambridge. A Commission
has been inquiring into the revenues of the many
wealthy societies, in more or less direct connection
with the Universities, resident in those towns. It
is said that the Commission has reported, and
that, for the first time in recorded history, the
nation, and perhaps the Colleges themselves, will
know what they are worth. And it was announced
that a statesman, who, whatever his other merits
or defects, has aims above the level of mere party
fighting, and a clear vision into the most complex

practical problems, meant to deal with these revenues.

But, *Bos locutus est.* That mysterious independent variable of political calculation, Public Opinion—which some whisper is, in the present case, very much the same thing as publican's opinion—has willed otherwise. The Heads may return to their wonted slumbers—at any rate for a space.

Is the spirit of change, which is working thus vigorously in the South, likely to affect the Northern Universities, and if so, to what extent? The violence of fermentation depends, not so much on the quantity of the yeast, as on the composition of the wort, and its richness in fermentable material; and, as a preliminary to the discussion of this question, I venture to call to your minds the essential and fundamental differences between the Scottish and the English type of University.

Do not charge me with anything worse than official egotism, if I say that these differences appear to be largely symbolised by my own existence. There is no Rector in an English University. Now, the organisation of the members of a University into Nations, with their elective Rector, is the last relic of the primitive constitution of Universities. The Rectorate was the most important of all offices in that University of Paris, upon the model of which the University

is now part of Prussia, objected to the Frankish
king's measures; no doubt the priests, who had
never hesitated about sacrificing all unbelievers in
their fantastic deities and futile conjurations, were
the loudest in chanting the virtues of toleration;
no doubt they denounced as a cruel persecutor
the man who would not allow them, however
sincere they might be, to go on spreading de-
lusions which debased the intellect, as much as
they deadened the moral sense, and undermined
the bonds of civil allegiance; no doubt, if they
had lived in these times, they would have been
able to show, with ease, that the king's proceedings were totally contrary to the best liberal
principles.   But it may be said, in justification of
the Teutonic ruler, first, that he was born before
those principles, and did not suspect that the best
way of getting disorder into order was to let it
alone; and, secondly, that his rough and questionable proceedings did, more or less, bring about the
end he had in view.   For, in a couple of centuries,
the schools he sowed broadcast produced their
crop of men, thirsting for knowledge and craving
for culture.   Such men gravitating towards Paris,
as a light amidst the darkness of evil days, from
Germany, from Spain, from Britain, and from
Scandinavia, came together by natural affinity.
By degrees they banded themselves into a society,
which, as its end was the knowledge of all things
knowable, called itself a " *Studium   Generale;* "

and when it had grown into a recognised corporation, acquired the name of " *Universitas Studii
Generalis*," which, mark you, means not a " Useful
Knowledge Society," but a " Knowledge-of-things-
in-general Society."

And thus the first " University," at any rate on
this side of the Alps, came into being. Originally
it had but one Faculty, that of Arts. Its aim was
to be a centre of knowledge and culture ; not to
be, in any sense, a technical school.

The scholars seem to have studied Grammar,
Logic, and Rhetoric ; Arithmetic and Geometry ;
Astronomy ; Theology ; and Music. Thus, their
work, however imperfect and faulty, judged by
modern lights, it may have been, brought them
face to face with all the leading aspects of the
many-sided mind of man. For these studies did
really contain, at any rate in embryo—sometimes,
it may be, in caricature—what we now call
Philosophy, Mathematical and Physical Science,
and Art. And I doubt if the curriculum of any
modern University shows so clear and generous a
comprehension of what is meant by culture, as
this old Trivium and Quadrivium does.

The students who had passed through the
University course, and had proved themselves
competent to teach, became masters and teachers
of their younger brethren. Whence the distinction of Masters and Regents on the one hand, and
Scholars on the other.

Rapid growth necessitated organisation. The Masters and Scholars of various tongues and countries grouped themselves into four Nations; and the Nations, by their own votes at first, and subsequently by those of their Procurators, or representatives, elected their supreme head and governor, the Rector—at that time the sole representative of the University, and a very real power, who could defy Provosts interfering from without; or could inflict even corporal punishment on disobedient members within the University.

Such was the primitive constitution of the University of Paris. It is in reference to this original state of things that I have spoken of the Rectorate, and all that appertains to it, as the sole relic of that constitution.

But this original organisation did not last long. Society was not then, any more than it is now, patient of culture, as such. It says to everything, "Be useful to me, or away with you." And to the learned, the unlearned man said then, as he does now, " What is the use of all your learning, unless you can tell me what I want to know ? I am here blindly groping about, and constantly damaging myself by collision with three mighty powers, the power of the invisible God, the power of my fellow Man, and the power of brute Nature. Let your learning be turned to the study of these powers, that I may know how I am to comport myself with regard to them." In answer to this

demand, some of the Masters of the Faculty of
Arts devoted themselves to the study of Theology,
some to that of Law, and some to that of
Medicine ; and they became Doctors—men learned
in those technical, or, as we now call them, professional, branches of knowledge. Like cleaving
to like, the Doctors formed schools, or Faculties,
of Theology, Law, and Medicine, which sometimes
assumed airs of superiority over their parent, the
Faculty of Arts, though the latter always asserted
and maintained its fundamental supremacy.

The Faculties arose by process of natural
differentiation out of the primitive University.
Other constituents, foreign to its nature, were
speedily grafted upon it. One of these extraneous
elements was forced into it by the Roman Church,
which in those days asserted with effect, that
which it now asserts, happily without any effect
in these realms, its right of censorship and
control over all teaching. The local habitation
of the University lay partly in the lands attached
to the monastery of S. Geneviève, partly in the
diocese of the Bishop of Paris ; and he who would
teach must have the licence of the Abbot, or of
the Bishop, as the nearest representative of the
Pope, so to do, which licence was granted by the
Chancellors of these Ecclesiastics.

Thus, if I am what archæologists call a
"survival" of the primitive head and ruler of the
University, your Chancellor stands in the same

relation to the Papacy; and, with all respect for his Grace, I think I may say that we both look terribly shrunken when compared with our great originals.

Not so is it with a second foreign element, which silently dropped into the soil of Universities, like the grain of mustard-seed in the parable; and, like that grain, grew into a tree, in whose branches a whole aviary of fowls took shelter. That element is the element of Endowment. It differed from the preceding, in its original design to serve as a prop to the young plant, not to be a parasite upon it. The charitable' and the humane, blessed with wealth, were very early penetrated by the misery of the poor student. And the wise saw that intellectual ability is not so common or so unimportant a gift that it should be allowed to run to waste upon mere handicrafts and chares. The man who was a blessing to his contemporaries, but who so often has been converted into a curse, by the blind adherence of his posterity to the letter, rather than to the spirit, of his wishes—I mean the "pious founder"—gave money and lands, that the student, who was rich in brain and poor in all else, might be taken from the plough or from the stithy, and enabled to devote himself to the higher service of mankind; and built colleges and halls in which he might be not only housed and fed, but taught.

The Colleges were very generally placed in

strict subordination to the University by their founders; but, in many cases, their endowment, consisting of land, has undergone an "unearned increment," which has given these societies a continually increasing weight and importance as against the unendowed, or fixedly endowed, University. In Pharaoh's dream, the seven lean kine eat up the seven fat ones. In the reality of historical fact, the fat Colleges have eaten up the lean Universities.

Even here in Aberdeen, though the causes at work may have been somewhat different, the effects have been similar ; and you see how much more substantial an entity is the Very Reverend the Principal, analogue, if not homologue, of the Principals of King's College, than the Rector, lineal representative of the ancient monarchs of the University, though now, little more than a "king of shreds and patches."

Do not suppose that, in thus briefly tracing the process of University metamorphosis, I have had any intention of quarrelling with its results. Practically, it seems to me that the broad changes effected in 1858 have given the Scottish Universities a very liberal constitution, with as much real approximation to the primitive state of things as is at all desirable. If your fat kine have eaten the lean, they have not lain down to chew the cud ever since. The Scottish Universities, like the English, have diverged widely enough from their

primitive model; but I cannot help thinking that
the northern form has remained more faithful to
its original, not only in constitution, but, what is
more to the purpose, in view of the cry for change,
in the practical application of the endowments
connected with it.

In Aberdeen, these endowments are numerous,
but so small that, taken altogether, they are not
equal to the revenue of a single third-rate English
college. They are scholarships, not fellowships;
aids to do work—not rewards for such work as it
lies within the reach of an ordinary, or even an
extraordinary, young man to do. You do not
think that passing a respectable examination is a
fair equivalent for an income, such as many a
grey-headed veteran, or clergyman would envy;
and which is larger than the endowment of many
Regius chairs. You do not care to make your
University a school of manners for the rich; of
sports for the athletic; or a hot-bed of high-fed,
hypercritical refinement, more destructive to vigour
and originality than are starvation and oppression.
No; your little Bursaries of ten and twenty (I
believe even fifty) pounds a year, enabled any boy
who has shown ability in the course of his education
in those remarkable primary schools, which have
made Scotland the power she is, to obtain the
highest culture the country can give him; and
when he is armed and equipped, his Spartan
Alma Mater tells him that, so far, he has had his

wages for his work, and that he may go and earn the rest.

When I think of the host of pleasant, moneyed, well-bred young gentlemen, who do a little learning and much boating by Cam and Isis, the vision is a pleasant one ; and, as a patriot, I rejoice that the youth of the upper and richer classes of the nation receive a wholesome and a manly training, however small may be the modicum of knowledge they gather, in the intervals of this, their serious business.   I admit, to the full, the social and political value of that training.   But, when I proceed to consider that these young men may be said to represent the great bulk of what the Colleges have to show for their enormous wealth, plus, at least, a hundred and fifty pounds a year apiece which each undergraduate costs his parents or guardians, I feel inclined to ask, whether the rate-in-aid of the education of the wealthy and professional classes, thus levied on the resources of the community, is not, after all, a little heavy ? And, still further, I am tempted to inquire what has become of the indigent scholars, the sons of the masses of the people whose daily labour just suffices to meet their daily wants, for whose benefit these rich foundations were largely, if not mainly, instituted ?   It seems as if Pharaoh's dream had been rigorously carried out, and that even the fat scholar has eaten the lean one.   And when I turn from this picture to the no less real vision of many a brave and frugal Scotch boy, spending his summer in hard manual labour, that he may have the privilege of wending his way in autumn to this University, with a bag of oatmeal, ten pounds in his pocket, and his own stout heart to depend upon through the northern winter; not bent on seeking

"The bubble reputation at the cannon's mouth,"

but determined to wring knowledge from the hard hands of penury; when I see him win through all such outward obstacles to positions of wide usefulness and well-earned fame; I cannot but think that, in essence, Aberdeen has departed but little from the primitive intention of the founders of Universities, and that the spirit of reform has so much to do on the other side of the Border, that it may be long before he has leisure to look this way.

As compared with other actual Universities, then, Aberdeen, may, perhaps, be well satisfied with itself. But do not think me an impracticable dreamer, if I ask you not to rest and be thankful in this state of satisfaction; if I ask you to consider awhile, how this actual good stands related to that ideal better, towards which both men and institutions must progress, if they would not retrograde.

In an ideal University, as I conceive it, a man should be able to obtain instruction in all forms

of knowledge, and discipline in the use of all the
methods by which knowledge is obtained.   In
such a University, the force of living example
should fire the student with a noble ambition to
emulate the learning of learned men, and to follow
in the footsteps of the explorers of new fields of
knowledge.   And the very air he breathes should
be charged with that enthusiasm for truth, that
fanaticism of veracity, which is a greater possession
than much learning; a nobler gift than the power
of increasing knowledge; by so much greater and
nobler than these, as the moral nature of man is
greater than the intellectual; for veracity is the
heart of morality.

But the man who is all morality and intellect,
although he may be good and even great, is, after
all, only half a man.   There is beauty in the
moral world and in the intellectual world; but
there is also a beauty which is neither moral nor
intellectual—the beauty of the world of Art.
There are men who are devoid of the power of
seeing it, as there are men who are born deaf and
blind, and the loss of those, as of these, is simply
infinite.   There are others in whom it is an over-
powering passion; happy men, born with the productive, or at lowest, the appreciative, genius of
the Artist.   But, in the mass of mankind, the
Æsthetic faculty, like the reasoning power and
the moral sense, needs to be roused, directed, and
cultivated; and I know not why the develop-

ment of that side of his nature, through which
man has access to a perennial spring of ennobling pleasure, should be omitted from any
comprehensive scheme of University education.

All Universities recognise Literature in the
sense of the old Rhetoric, which is art incarnate in
words.   Some, to their credit, recognise Art in its
narrower sense, to a certain extent, and confer
degrees for proficiency in some of its branches.   If
there are Doctors of Music, why should there be
no Masters of painting, of Sculpture, of Architecture ?   I should like to see Professors of the Fine
Arts in every University ; and instruction in some
branch of their work made a part of the Arts
curriculum.

I just now expressed the opinion that, in our
ideal University, a man should be able to obtain
instruction in all forms of knowledge.   Now, by
" forms of knowledge " I mean the great classes of
things knowable ; of which the first, in logical,
though not in natural, order is knowledge relating to
the scope and limits of the mental faculties of man,.
a form of knowledge which, in its positive aspect,
answers pretty much to Logic and part of
Psychology, while, on its negative and critical side,
it corresponds with Metaphysics.

A second class comprehends all that knowledge
which relates to man's welfare, so far as it is determined by his own acts, or what we call his conduct.   It answers to Moral and Religious philos-

ophy. Practically, it is the most directly valuable of all forms of knowledge, but speculatively, it is limited and criticised by that which precedes and by that which follows it in my order of enumeration.

A third class embraces knowledge of the phænomena of the Universe, as that which lies about the individual man ; and of the rules which those phænomena are observed to follow in the order of their occurrence, which we term the laws of Nature.

This is what ought to be called Natural Science, or Physiology, though those terms are hopelessly diverted from such a meaning; and it includes all exact knowledge of natural fact, whether Mathematical, Physical, Biological, or Social.

Kant has said that the ultimate object of all knowledge is to give replies to these three questions : What can I do ? What ought I to do ? What may I hope for ? The forms of knowledge which I have enumerated, should furnish such replies as are within human reach, to the first and second of these questions. While to the third, perhaps the wisest answer is, " Do what you can to do what you ought, and leave hoping and fearing alone."

If this be a just and an exhaustive classification of the forms of knowledge, no question as to their relative importance, or as to the superiority of one to the other, can be seriously raised.

On the face of the matter, it is absurd to ask
whether it is more important to know the limits of
one's powers; or the ends for which they ought to
be exerted; or the conditions under which they
must be exerted. One may as well inquire which
of the terms of a Rule of Three sum one ought to
know, in order to get a trustworthy result. Practical life is such a sum, in which your duty multiplied into your capacity, and divided by your
circumstances, gives you the fourth term in the
proportion, which is your deserts, with great
accuracy. All agree, I take it, that men ought
to have these three kinds of knowledge. The so-
called " conflict of studies " turns upon the question of how they may best be obtained.

The founders of Universities held the theory
that the Scriptures and Aristotle taken together,
the latter being limited by the former, contained
all knowledge worth having, and that the business
of philosophy was to interpret and co-ordinate
these two. I imagine that in the twelfth century
this was a very fair conclusion from known facts.
Nowhere in the world, in those days, was there
such an encyclopædia of knowledge of all three
classes, as is to be found in those writings. The
scholastic philosophy is a wonderful monument of
the patience and ingenuity with which the human
mind toiled to build up a logically consistent
theory of the Universe, out of such materials.
And that philosophy is by no means dead and

buried, as many vainly suppose. On the contrary, numbers of men of no mean learning and accomplishment, and sometimes of rare power and subtlety of thought, hold by it as the best theory of things which has yet been stated. And, what is still more remarkable, men who speak the language of modern philosophy, nevertheless think the thoughts of the schoolmen. " The voice is the voice of Jacob, but the hands are the hands of Esau." Every day I hear "Cause," "Law," " Force," " Vitality," spoken of as entities, by people who can enjoy Swift's joke about the meat-roasting quality of the smoke-jack, and comfort themselves with the reflection that they are not even as those benighted schoolmen.

Well, this great system had its day, and then it was sapped and mined by two influences. The first was the study of classical literature, which familiarised men with methods of philosophising; with conceptions of the highest Good ; with ideas of the order of Nature ; with notions of Literary and Historical Criticism; and, above all, with visions of Art, of a kind which not only would not fit into the scholastic scheme, but showed them a pre-Christian, and indeed altogether un-Christian world, of such grandeur and beauty that they ceased to think of any other. They were as men who had kissed the Fairy Queen, and wandering with her in the dim loveliness of the under-world,

cared not to return to the familiar ways of home
and fatherland, though they lay, at arm's length,
overhead.   Cardinals were more familiar with
Virgil than with Isaiah ; and Popes laboured, with
great success, to re-paganise Rome.

The second influence was the slow, but sure,
growth of the physical sciences.   It was discovered
that some results of speculative thought, of immense practical and theoretical importance, can be
verified by observation ; and are always true, however severely they may be tested.   Here, at any
rate, was knowledge, to the certainty of which no
authority could add, or take away, one jot or tittle,
and to which the tradition of a thousand years
was as insignificant as the hearsay of yesterday.
To the scholastic system, the study of classical
literature might be inconvenient and distracting,
but it was possible to hope that it could be kept
within bounds.   Physical science, on the other
hand, was an irreconcilable enemy, to be excluded
at all hazards.   The College of Cardinals has not
distinguished itself in Physics or Physiology ; and
no Pope has, as yet, set up public laboratories in
the Vatican.

People do not always formulate the beliefs on
which they act.   The instinct of fear and dislike
is quicker than the reasoning process; and I
suspect that, taken in conjunction with some
other causes, such instinctive aversion is at the

bottom of the long exclusion of any serious discipline in the physical sciences from the general curriculum of Universities ; while, on the other hand, classical literature has been gradually made the backbone of the Arts course.

I am ashamed to repeat here what I have said elsewhere, in season and out of season, respecting the value of Science as knowledge and discipline. But the other day I met with some passages in the Address to another Scottish University, of a great thinker, recently lost to us, which express so fully and yet so tersely, the truth in this matter that I am fain to quote them :—

"To question all things ;—never to turn away from any difficulty ; to accept no doctrine either from ourselves or from other people without a rigid scrutiny by negative criticism ; letting no fallacy, or incoherence, or confusion of thought, step by unperceived ; above all, to insist upon having the meaning of a word clearly understood before using it, and the meaning of a proposition before assenting to it ;—these are the lessons we learn " from workers-in Science. " With all this vigorous management of the negative element, they inspire no scepticism about the reality of truth or in- difference to its pursuit. The noblest enthusiasm, both for the search after truth and for applying it to its highest uses, pervades those writers." " In cultivating, therefore," science as an essential ingredient in education, " we are all the while laying an admirable foundation for ethical and philosophical culture." [1]

The passages I have quoted were uttered by John Stuart Mill; but you cannot hear inverted commas, and it is therefore right that I should add, without delay, that I have taken the liberty of substituting "workers in science" for "ancient dialecticians," and "Science as an essential ingredient in education" for "the ancient languages as our best literary education." Mill did, in fact, deliver a noble panegyric upon classical studies. I do not doubt its justice, nor presume to question its wisdom. But I venture to maintain that no wise or just judge, who has a knowledge of the facts, will hesitate to say that it applies with equal force to scientific training.

But it is only fair to the Scottish Universities to point out that they have long understood the value of Science as a branch of general education. I observe, with the greatest satisfaction, that candidates for the degree of Master of Arts in this University are required to have a knowledge, not only of Mental and Moral Philosophy, and of Mathematics and Natural Philosophy, but of Natural History, in addition to the ordinary Latin and Greek course; and that a candidate may take honours in these subjects and in Chemistry.

[1] Inaugural Address delivered to the University of St. Andrew, February 1, 1867, by J. S. Mill, Rector of the University (pp. 32, 33).

I do not know what the requirements of your examiners may be, but I sincerely trust they are not satisfied with a mere book knowledge of these matters. For my own part I would not raise a finger, if I could thereby introduce mere book work in science into every Arts curriculum in the country. Let those who want to study books devote themselves to Literature, in which we have the perfection of books, both as to substance and as to form. If I may paraphrase Hobbes's well-known aphorism, I would say that "books are the money of Literature, but only the counters of Science," Science (in the sense in which I now use the term) being the knowledge of fact, of which every verbal description is but an incomplete and symbolic expression. And be assured that no teaching of science is worth anything, as a mental discipline, which is not based upon direct perception of the facts, and practical exercise of the observing and logical faculties upon them. Even in such a simple matter as the mere comprehension of form, ask the most practised and widely informed anatomist what is the difference between his knowledge of a structure which he has read about, and his knowledge of the same structure when he has seen it for himself; and he will tell you that the two things are not comparable—the difference is infinite. Thus I am very strongly inclined to agree with some learned schoolmasters who say that, in their experience, the teaching of science is all waste time.

English University men remain in their present state of barbarous ignorance of even the rudiments of scientific culture.

Yet another step needs to be made before Science can be said to have taken its proper place in the Universities. That is its recognition as a Faculty, or branch of study demanding recognition and special organisation, on account of its bearing on the wants of mankind. The Faculties of Theology, Law, and Medicine, are technical schools, intended to equip men who have received general culture, with the special knowledge which is needed for the proper performance of the duties of clergymen, lawyers, and medical practitioners.

When the material well-being of the country depended upon rude pasture and agriculture, and still ruder mining ; in the days when all the innumerable applications of the principles of physical science to practical purposes were non-existent even as dreams ; days which men living may have heard their fathers speak of ; what little physical science could be seen to bear directly upon human life, lay within the province of Medicine. Medicine was the foster-mother of Chemistry, because it has to do with the preparation of drugs and the detection of poisons ; of Botany, because it enabled the physician to recognise medicinal herbs ; of Comparative Anatomy and Physiology, because the man who studied Human Anatomy and Physiology for purely medical purposes was led to extend his studies to the rest of the animal world.

Within my recollection, the only way in which a student could obtain anything like a training in Physical Science, was by attending the lectures of the Professors of Physical and Natural Science attached to the Medical Schools. But, in the course of the last thirty years, both foster-mother and child have grown so big, that they threaten not only to crush one another, but to press the very life out of the unhappy student who enters the nursery ; to the great detriment of all three.

I speak in the presence of those who know practically what medical education is ; for I may assume that a large proportion of my hearers are more or less advanced students of medicine. I appeal to the most industrious and conscientious among you, to those who are most deeply penetrated with a sense of the extremely serious responsibilities which attach to the calling of a medical practitioner, when I ask whether, out of the four years which you devote to your studies, you ought to spare even so much as an hour for any work which does not tend directly to fit you for your duties ?

Consider what that work is. Its foundation is a sound and practical acquaintance with the structure of the human organism, and with the modes and conditions of its action in health. I say a sound and practical acquaintance, to guard against the supposition that my intention is to suggest that you ought all to be minute anatomists and accomplished physiologists.· The devotion of your whole four years to Anatomy and Physiology alone, would be totally insufficient to attain that end. What I mean is, the sort of practical, familiar, finger-end knowledge which a watchmaker has of a watch, and which you expect that craftsman, as an honest man, to have, when you entrust a watch that goes badly, to him. It is a kind of knowledge which is to be acquired, not in the lecture-room, nor in the library, but in the dissecting-room and the laboratory. It is to be had not by sharing your attention between these and sundry other subjects, but by concentrating your minds, week after week, and month after month, six or seven hours a day, upon all the complexities of organ and function, until each of the greater truths of anatomy and physiology has become an organic part of your minds—until you would know them if you were roused and questioned in the middle of the night, as a man knows the geography of his native place and the daily life of his home. That is the sort of knowledge which, once obtained, is a life-long possession. Other occupations may fill your minds—it may grow dim, and seem to be forgotten—but there it is, like the inscription on a battered and defaced coin, which comes out when you warm it.

If I had the power to remodel Medical Education, the first two years of the medical curriculum should be devoted to nothing but such thorough study of Anatomy and Physiology, with Physiological Chemistry and Physics; the student should then pass a real, practical examination in these subjects; and, having gone through that ordeal satisfactorily, he should be troubled no more with them. His whole mind should then be given with equal intentness to Therapeutics, in its broadest sense, to Practical Medicine and to Surgery, with instruction in Hygiene and in Medical Jurisprudence; and of these subjects only —surely there are enough of them—should he be required to show a knowledge in his final examination.

I cannot claim any special property in this theory of what the medical curriculum should be, for I find that views, more or less closely approximating these, are held by all who have seriously considered the very grave and pressing question of Medical Reform; and have, indeed, been carried into practice, to some extent, by the most enlightened Examining Boards. I have heard but two kinds of objections to them. There is first, the objection of vested interests, which I will not deal with here, because I want to make myself as pleasant as I can, and no discussions are so unpleasant as those which turn on such points. And there is, secondly, the much more respectable

objection, which takes the general form of the
reproach that, in thus limiting the curriculum, we
are seeking to narrow it. We are told that the
medical man ought to be a person of good education and general information, if his profession is to
hold its own among other professions ; that he
ought to know Botany, or else, if he goes abroad,
he will not be able to tell poisonous fruits from
edible ones ; that he ought to know drugs, as a
druggist knows them, or he will not be able to tell
sham bark and senna from the real articles ; that
he ought to know Zoology, because—well, I really
have never been able to learn exactly why he is to
be expected to know zoology. There is, indeed,
a popular superstition, that doctors know all
about things that are queer or nasty to the general
mind, and may, therefore, be reasonably expected
to know the "barbarous binomials" applicable to
snakes, snails, and slugs ; an amount of information with which the general mind is usually completely satisfied. And there is a scientific superstition that Physiology is largely aided by
Comparative Anatomy—a superstition which, like
most superstitions, once had a grain of truth at
bottom ; but the grain has become homœopathic,
since Physiology took its modern experimental
development, and became what it is now, the application of the principles of Physics and Chemistry
to the elucidation of the phænomena of life.

I hold as strongly as any one can do, that the

"If a man could be sure
That his life would endure
For the space of a thousand long years——"

he might do a number of things not practicable
under present conditions. Methuselah might, with
much propriety, have taken half a century to get
his doctor's degree ; and might, very fairly, have
been required to pass a practical examination upon
the contents of the British Museum, before commencing practice as a promising young fellow of
two hundred, or thereabouts. But you have four
years to do your work in, and are turned loose, to
save or slay, at two or three and twenty.

Now, I put it to you, whether you think that,
when you come down to the realities of life—when
you stand by the sick-bed, racking you brains for
the principles which shall furnish you with the
means of interpreting symptoms, and forming a
rational theory of the condition of your patient, it
will be satisfactory for you to find that those
principles are not there—although, to use the
examination slang which is unfortunately too
familiar to me, you can quite easily "give an
account of the leading peculiarities of the *Marsu-
pialia*," or "enumerate the chief characters of the
Compositæ," or "state the class and order of the
animal from which Castoreum is obtained."

I really do not think that state of things will
be satisfactory to you ; I am very sure it will not
be so to your patient. Indeed, I am so narrow-

minded myself, that if I had to choose between
two physicians—one who did not know whether a
whale is a fish or not, and could not tell gentian
from ginger, but did understand the applications of
the institutes of medicine to his art; while the
other, like Talleyrand's doctor, "knew everything,
even a little physic"—with all my love for
breadth of culture, I should assuredly consult the
former.

It is not pleasant to incur the suspicion of an
inclination to injure or depreciate particular
branches of knowledge. But the fact that one of
those which I should have no hesitation in ex-
cluding from the medical curriculum, is that to
which my own life has been specially devoted,
should, at any rate, defend me from the suspicion
of being urged to this course by any but the very
gravest considerations of the public welfare.

And I should like, further, to call your attention
to the important circumstance that, in thus proposing the exclusion of the study of such branches
of knowledge as Zoology and Botany, from those
compulsory upon the medical student, I am not,
for a moment, suggesting their exclusion from the
University. I think that sound and practical
instruction in the elementary facts and broad
principles of Biology should form part of the Arts
Curriculum: and here, happily, my theory is in
entire accordance with your practice. Moreover,
as I have already said, I have no sort of doubt

that, in view of the relation of Physical Science to the practical life of the present day, it has the same right as Theology, Law, and Medicine, to a Faculty of its own in which men shall be trained to be professional men of science. It may be doubted whether Universities are the places for technical schools of Engineering or applied Chemistry, or Agriculture. But there can surely be little question, that instruction in the branches of Science which lie at the foundation of these Arts, of a far more advanced and special character than could, with any propriety, be included in the ordinary Arts Curriculum, ought to be obtainable by means of a duly organised Faculty of Science in every University.

The establishment of such a Faculty would have the additional advantage of providing, in some measure, for one of the greatest wants of our time and country. I mean the proper support and encouragement of original research.

The other day, an emphatic friend of mine committed himself to the opinion that, in England, it is better for a man's worldly prospects to be a drunkard, than to be smitten with the divine dipsomania of the original investigator. I am inclined to think he was not far wrong. And, be it observed, that the question is not, whether such a man shall be able to make as much out of his abilities as his brother, of like ability, who goes into Law, or Engineering, or Commerce ; it is not a question of

" maintaining a due number of saddle horses," as
George Eliot somewhere puts it—it is a question
of living or starving.

If a student of my own subject shows power and
originality, I dare not advise him to adopt a
scientific career; for, supposing he is able to
maintain himself until he has attained distinction,
I cannot give him the assurance that any amount
of proficiency in the Biological Sciences will be
convertible into, even the most modest, bread and
cheese.   And I believe that the case is as bad, or
perhaps worse, with other branches of Science.
In this respect Britain, whose immense wealth
and prosperity hang upon the thread of Applied
Science, is far behind France, and infinitely behind
Germany.

And the worst of it is, that it is very difficult to
see one's way to any immediate remedy for this
state of affairs which shall be free from a tendency
to become worse than the disease.

Great schemes for the Endowment of Research
have been proposed.   It has been suggested, that
Laboratories for all branches of Physical Science,
provided with every apparatus needed by the investigator, shall be established by the State : and
shall be accessible, under due conditions and
regulations, to all properly qualified persons.   I
see no objection to the principle of such a proposal.
If it be legitimate to spend great sums of money
on public Libraries and public collections of Painting

and Sculpture, in aid of the Man of Letters, or the Artist, or for the mere sake of affording pleasure to the general public. I apprehend that it cannot be illegitimate to do as much for the promotion of scientific investigation. To take the lowest ground, as a mere investment of money, the latter is likely to be much more immediately profitable. To my mind, the difficulty in the way of such schemes is not theoretical, but practical. Given the laboratories, how are the investigators to be maintained ? What career is open to those who have been thus encouraged to leave bread-winning pursuits ? If they are to be provided for by endowment, we come back to the College Fellowship system, the results of which, for Literature, have not been so brilliant that one would wish to see it extended to Science ; unless some much better securities than at present exist can be taken that it will foster real work. You know that among the Bees, it depends on the kind of cell in which the egg is deposited, and the quantity and quality of food which is supplied to the grub, whether it shall turn out a busy little worker or a big idle queen. And, in the human hive, the cells of the endowed larvæ are always tending to enlarge, and their food to improve, until we get queens, beautiful to behold, but which gather no honey and build no comb.

I do not say that these difficulties may not be overcome, but their gravity is not to be lightly estimated.

In the meanwhile, there is one step in the direction of the endowment of research which is free from such objections. It is possible to place the scientific enquirer in a position in which he shall have ample leisure and opportunity for original work, and yet shall give a fair and tangible equivalent for those privileges. The establishment of a Faculty of Science in every University, implies that of a corresponding number of Professorial chairs, the incumbents of which need not be so burdened with teaching as to deprive them of ample leisure for original work. I do not think that it is any impediment to an original investigator to have to devote a moderate portion of his time to lecturing, or superintending practical instruction. On the contrary, I think it may be, and often is, a benefit to be obliged to take a comprehensive survey of your subject; or to bring your results to a point, and give them, as it were, a tangible objective existence. The besetting sins of the investigator are two : the one is the desire to put aside a subject, the general bearings of which he has mastered himself, and pass on to something which has the attraction of novelty; and the other, the desire for too much perfection, which leads him to

> " Add and alter many times,
> Till all be ripe and rotten ; "

to spend the energies which should be reserved for action in whitening the decks and polishing the guns.

The obligation to produce results for the instruction of others, seems to me to be a more
effectual check on these tendencies than even the
love of usefulness or the ambition for fame.
But supposing the Professorial forces of our
University to be duly organised, there remains an
important question, relating to the teaching power,
to be considered. Is the Professorial system—the
system, I mean, of teaching in the lecture-room
alone, and leaving the student to find his own way
when he is outside the lecture-room—adequate to
the wants of learners ?   In answering this question, I confine myself to my own province, and I
venture to reply for Physical Science, assuredly
and undoubtedly, No. As I have already intimated,
practical work in the Laboratory is absolutely
indispensable, and that practical work must be
guided and superintended by a sufficient staff of
Demonstrators, who are for Science what Tutors
are for other branches of study. And there must
be a good supply of such Demonstrators. I doubt
if the practical work of more than twenty students
can be properly superintended by one Demon-
strator. If we take the working day at six hours,
that is less than twenty minutes apiece—not a
very large allowance of time for helping a dull
man, for correcting an inaccurate one, or even for
making an intelligent student clearly apprehend
what he is about. And, no doubt, the supplying
of a proper amount of this tutorial, practical

teaching, is a difficulty in the way of giving proper
instruction in Physical Science in such Universities as that of Aberdeen, which are devoid of
endowments; and, unlike the English Universities,
have no moral claim on the funds of richly
endowed bodies to supply their wants.

Examination—thorough, searching examination
—is an indispensable accompaniment of teaching;
but I am almost inclined to commit myself to the
very heterodox proposition that it is a necessary
evil. I am a very old Examiner, having, for some
twenty years past, been occupied with examinations
on a considerable scale, of all sorts and conditions
of men, and women too,—from the boys and girls
of elementary schools to the candidates for Honours
and Fellowships in the Universities. I will not
say that, in this case as in so many others, the
adage, that familiarity breeds contempt, holds
good; but my admiration for the existing system
of examination and its products, does not wax
warmer as I see more of it.   Examination, like
fire, is a good servant, but a bad master; and there
seems to me to be some danger of its becoming our
master. I by no means stand alone in this opinion.
Experienced friends of mine do not hesitate to say
that students whose career they watch, appear to
them to become deteriorated by the constant effort
to pass this or that examination, just as we hear of
men's brains becoming affected by the daily necessity of catching a train.   They work to pass, not

to know ; and outraged Science takes her revenge.
They do pass, and they don't know. I have passed
sundry examinations in my time, not without
credit, and I confess I am ashamed to think how
very little real knowledge underlay the torrent of
stuff which I was able to pour out on paper. In
fact, that which examination, as ordinarily conducted, tests, is simply a man's power of work
under stimulus, and his capacity for rapidly and
clearly producing that which, for the time, he has
got into his mind. Now, these faculties are by no
means to be despised. They are of great value in
practical life, and are the making of many an
advocate, and of many a so-called statesman. But
in the pursuit of truth, scientific or other, they
count for very little, unless they are supplemented
by that long-continued, patient " intending of the
mind," as Newton phrased it, which makes very
little show in Examinations. I imagine that an
Examiner who knows his students personally, must
not unfrequently have found himself in the position of finding A's paper better than B's, though
his own judgment tells him, quite clearly, that B
is the man who has the larger share of genuine
capacity.

Again, there is a fallacy about Examiners. It
is commonly supposed that any one who knows a
subject is competent to teach it ; and no one seems
to doubt that any one who knows a subject is
competent to examine in it. I believe both these

opinions to be serious mistakes: the latter, perhaps, the more serious of the two.   In the first
place, I do not believe that any one who is not,
or has not been, a teacher is really qualified to
examine advanced students.   And in the second
place, Examination is an Art, and a difficult one,
which has to be learned like all other arts.

Beginners always set too difficult questions—
partly because they are afraid of being suspected
of ignorance if they set easy ones, and partly
from not understanding their business.   Suppose
that you want to test the relative physical
strength of a score of young men.   You do not
put a hundredweight down before them, and tell
each to swing it round.   If you do, half of them
won't be able to lift it at all, and only one or two
will be able to perform the task.   You must give
them half a hundredweight, and see how they
manœuvre that, if you want to form any estimate
of the muscular strength of each.   So, a practised
Examiner will seek for information respecting the
mental vigour and training of candidates from the
way in which they deal with questions easy
enough to let reason, memory, and method have
free play.

No doubt, a great deal is to be done by the
careful selection of Examiners, and by the copious
introduction of practical work, to remove the evils
inseparable from examination; but, under the
best of circumstances, I believe that examination

will remain but an imperfect test of knowledge,
and a still more imperfect test of capacity, while
it tells next to nothing about a man's power as an
investigator.

There is much to be said in favour of restricting
the highest degrees in each Faculty, to those who
have shown evidence of such original power, by
prosecuting a research under the eye of the
Professor in whose province it lies; or, at any
rate, under conditions which shall afford satisfactory proof that the work is theirs. The notion
may sound revolutionary, but it is really very
old; for, I take it, that it lies at the bottom of
that presentation of a thesis by the candidate for
a doctorate, which has now, too often, become
little better than a matter of form.

Thus far, I have endeavoured to lay before
you, in a too brief and imperfect manner, my
views respecting the teaching half—the Magistri
and Regentes—of the University of the Future.
Now let me turn to the learning half—the
Scholares.

If the Universities are to be the sanctuaries of
the highest culture of the country, those who
would enter that sanctuary must not come with
unwashed hands. If the good seed is to yield its
hundredfold harvest, it must not be scattered
amidst the stones of ignorance, or the tares of
undisciplined indolence and wantonness. On the

contrary, the soil must have been carefully prepared, and the Professor should find that the operations of clod-crushing, draining, and weeding, and even a good deal of planting, have been done by the Schoolmaster.

That is exactly what the Professor does not find in any University in the three Kingdoms that I can hear of—the reason of which state of things lies in the extremely faulty organisation of the majority of secondary schools.   Students come to the Universities ill-prepared in classics and mathematics, not at all prepared in anything else; and half their time is spent in learning that which they ought to have known when they came.

I sometimes hear it said that the Scottish Universities differ from the English, in being to a much greater extent places of comparatively elementary education for a younger class of students.   But it would seem doubtful if any great difference of this kind really exists ; for a high authority, himself Head of an English College, has solemnly affirmed that : "Elementary teaching of youths under twenty is now the only function performed by the University ; " and that Colleges are "boarding schools in which the elements of the learned languages are taught to youths." [1]

[1] *Suggestions for Academical Organisation, with Especial Reference to Oxford.*   By the Rector of Lincoln.

This is not the first time that I have quoted those remarkable assertions. I should like to engrave them in public view, for they have not been refuted ; and I am convinced that if their import is once clearly apprehended, they will play no mean part when the question of University reorganisation, with a view to practical measures, comes on for discussion. You are not responsible for this anomalous state of affairs now; but, as you pass into active life and acquire the political influence to which your education and your position should entitle you, you will become responsible for it, unless each in his sphere does his best to alter it, by insisting on the improvement of secondary schools.

Your present responsibility is of another, though not less serious, kind. Institutions do not make men, any more than organisation makes life ; and even the ideal University we have been dreaming about will be but a superior piece of mechanism, unless each student strive after the ideal of the Scholar. And that ideal, it seems to me, has never been better embodied than by the great Poet, who, though lapped in luxury, the favourite of a Court, and the idol of his countrymen, remained through all the length of his honoured years a Scholar in Art, in Science, and in Life.

"Wouldst shape a noble life ? Then cast
No backward glances towards the past :

by no conditions save these :—That the principal shall not be employed in building : that the funds shall be appropriated, in equal proportions, to the promotion of natural knowledge and to the alleviation of the bodily sufferings of mankind; and, finally, that neither political nor ecclesiastical sectarianism shall be permitted to disturb the impartial distribution of the testator's benefactions.

In my experience of life a truth which sounds very much like a paradox has often asserted itself: namely, that a man's worst difficulties begin when he is able to do as he likes. So long as a man is struggling with obstacles he has an excuse for failure or shortcoming; but when fortune removes them all and gives him the power of doing as he thinks best, then comes the time of trial. There is but one right, and the possibilities of wrong are infinite. I doubt not that the trustees of the Johns Hopkins University felt the full force of this truth when they entered on the administration of their trust a year and a half ago; and I can but admire the activity and resolution which have enabled them, aided by the able president whom they have selected, to lay down the great outlines of their plan, and carry it thus far into execution. It is impossible to study that plan without perceiving that great care, forethought, and sagacity, have been bestowed upon it, and that it demands the most

respectful consideration. I have been endeavour-
ing to ascertain how far the principles which
underlie it are in accordance with those which
have been established in my own mind by much
and long-continued thought upon educational
questions. Permit me to place before you the
result of my reflections.

Under one aspect a university is a particular
kind of educational institution, and the views
which we may take of the proper nature of a
university are corollaries from those which we
hold respecting education in general. I think it
must be admitted that the school should prepare
for the university, and that the university should
crown ·the edifice, the foundations of which are
laid in the school. University education should
not be something distinct from elementary education, but should be the natural outgrowth and
development of the latter. Now I have a very
clear conviction as to what elementary education
ought to be ; what it really may be, when properly
organised ; and what I think it will be, before
many years have passed over our heads, in England and in America. Such education should
enable an average boy of fifteen or sixteen to
read and write his own language with ease and
accuracy, and with a sense of literary excellence
derived from the study of our classic writers :
to have a general acquaintance with the history
of his own country and with the great laws of

social existence; to have acquired the rudiments
of the physical and psychological sciences, and
a fair knowledge of elementary arithmetic and
geometry. He should have obtained an acquaintance with logic rather by example than by precept;
while the acquirement of the elements of music
and drawing should have been pleasure rather
than work.

It may sound strange to many ears if I venture
to maintain the proposition that a young person,
educated thus far, has had a liberal, though perhaps not a full, education. But it seems to me
that such training as that to which I have referred may be termed liberal, in both the senses in
which that word is employed, with perfect accuracy.
In the first place, it is liberal in breadth. It
extends over the whole ground of things to be
known and of faculties to be trained, and it gives
equal importance to the two great sides of human
activity—art and science. In the second place,
it is liberal in the sense of being an education
fitted for free men; for men to whom every career
is open, and from whom their country may demand
that they should be fitted to perform the duties
of any career. I cannot too strongly impress
upon you the fact that, with such a primary education as this, and with no more than is to be
obtained by building strictly upon its lines, a man
of ability may become a great writer or speaker,
a statesman, a lawyer, a man of science, painter,

specialise the instruction in each department.
Thus literature and philology, represented in the
elementary school by English alone, in the university will extend over the ancient and modern
languages. History, which, like charity, best
begins at home, but, like charity, should not end
there, will ramify into anthropology, archæology,
political history, and geography, with the history
of the growth of the human mind and of its products in the shape of philosophy, science, and art.
And the university will present to the student
libraries, museums of antiquities, collections of
coins, and the ·like, which will efficiently subserve
these studies. Instruction in the elements of
social economy, a most essential, but hitherto
sadly-neglected part of elementary education, will
develop in the university into political economy,
sociology, and law. Physical science will have
its great divisions of physical geography, with
geology and astronomy ; physics ; chemistry and
biology ; represented not merely by professors and
their lectures, but by laboratories, in which the
students, under guidance of demonstrators, will
work out facts for themselves and come into that
direct contact with reality which constitutes the
fundamental distinction of scientific education.
Mathematics will soar into its highest regions;
while the high peaks of philosophy may be scaled
by those whose aptitude for abstract thought has
been awakened by elementary logic. Finally,

schools of pictorial and plastic art, of architecture, and of music, will offer a thorough discipline in the principles and practice of art to those in whom lies nascent the rare faculty of æsthetic representation, or the still rarer powers of creative genius.

The primary school and the university are the alpha and omega of education. Whether institutions intermediate between these (so-called secondary schools) should exist, appears to me to be a question of practical convenience. If such schools are established, the important thing is that they should be true intermediaries between the primary school and the university, keeping on the wide track of general culture, and not sacrificing one branch of knowledge for another.

Such appear to me to be the broad outlines of the relations which the university, regarded as a place of education, ought to bear to the school, but a number of points of detail require some consideration, however briefly and imperfectly I can deal with them. In the first place, there is the important question of the limitations which should be fixed to the entrance into the university ; or, what qualifications should be required of those who propose to take advantage of the higher training offered by the university. On the one hand, it is obviously desirable that the time and opportunities of the university should not be wasted

in conferring such elementary instruction as can
be obtained elsewhere; while, on the other hand,
it is no less desirable that the higher instruction
of the university should be made accessible to
every one who can take advantage of it, although
he may not have been able to go through any
very extended course of education. My own
feeling is distinctly against any absolute and defined
preliminary examination, the passing of which shall
be an essential condition of admission to the
university. I would admit to the university any one
who could be reasonably expected to profit by the
instruction offered to him; and I should be inclined,
on the whole, to test the fitness of the student,
not by examination before he enters the university,
but at the end of his first term of study. If, on
examination in the branches of knowledge to which
he has devoted himself, he show himself deficient
in industry or in capacity, it will be best for the
university and best for himself, to prevent him
from pursuing a vocation for which he is obviously
unfit. And I hardly know of any other method
than this by which his fitness or unfitness can be
safely ascertained, though no doubt a good deal may
be done, not by formal cut and dried examination,
but by judicious questioning, at the outset of his
career.

Another very important and difficult practical
question is, whether a definite course of study
shall be laid down for those who enter the

university; whether a curriculum shall be pre-
scribed; or whether the student shall be allowed
to range at will among the subjects which are
open to him. And this question is inseparably
connected with another, namely, the conferring of
degrees. It is obviously impossible that any
student should pass through the whole of the
series of courses of instruction offered by a
university. If a degree is to be conferred as a
mark of proficiency in knowledge, it must be
given on the ground that the candidate is pro-
ficient in a certain fraction of those studies; and
then will arise the necessity of insuring an equivalency of degrees, so that the course by which a
degree is obtained shall mark approximately an
equal amount of labour and of acquirements, in
all cases. But this equivalency can hardly be
secured in any other way than by prescribing a
series of definite lines of study. This is a matter
which will require grave consideration. The important points to bear in mind, I think, are that
there should not be too many subjects in the
curriculum, and that the aim should be the
attainment of thorough and sound knowledge of
each.

One half of the Johns Hopkins bequest is
devoted to the establishment of a hospital, and
it was the desire of the testator that the university and the hospital should co-operate in the
promotion of medical education. The trustees

will unquestionably take the best advice that is
to be had as to the construction and administration of the hospital. In respect to the former
point, they will doubtless remember that a
hospital may be so arranged as to kill more than
it cures ; and, in regard to the latter, that a
hospital may spread the spirit of pauperism
among the well-to-do, as well as relieve the
sufferings of the destitute. It is not for me to
speak on these topics—rather let me confine
myself to the one matter on which my experience
as a student of medicine, and an examiner of long
standing, who has taken a great interest in the
subject of medical education, may entitle me to
a hearing. I mean the nature of medical education itself, and the co-operation of the university
in its promotion.

What is the object of medical education? It
is to enable the practitioner, on the one hand, to
prevent disease by his knowledge of hygiene; on
the other hand, to divine its nature, and to
alleviate or cure it, by his knowledge of pathology,
therapeutics, and practical medicine. That is his
business in life, and if he has not a thorough and
practical knowledge of the conditions of health,
of the causes which tend to the establishment
of disease, of the meaning of symptoms, and of
the uses of medicines and operative appliances,
he is incompetent, even if he were the best
anatomist, or physiologist, or chemist, that ever

took a gold medal or won a prize certificate. This is one great truth respecting medical education. Another is, that all practice in medicine is based upon theory of some sort or other ; and therefore, that it is desirable to have such theory in the closest possible accordance with fact. The veriest empiric who gives a drug in one case because he has seen it do good in another of apparently the same sort, acts upon the theory that similarity of superficial symptoms means similarity of lesions ; which, by the way, is perhaps as wild an hypothesis as could be invented. To understand the nature of disease we must understand health, and the understanding of the healthy body means the having a knowledge of its structure and of the way in which its manifold actions are performed, which is what is technically termed human anatomy and human physiology. The physiologist again must needs possess an acquaintance with physics and chemistry, inasmuch as physiology is, to a great extent, applied physics and chemistry. For ordinary purposes a limited amount of such knowledge is all that is needful ; but for the pursuit of the higher branches of physiology no knowledge of these branches of science can be too extensive, or too profound. Again, what we call therapeutics, which has to do with the action of drugs and medicines on the living organism, is, strictly speaking, a branch of experimental physiology, and is daily receiving a greater and greater experimental development.

The third great fact which is to be taken into consideration in dealing with medical education, is that the practical necessities of life do not, as a rule, allow aspirants to medical practice to give more than three, or it may be four years to their studies. Let us put it at four years, and then reflect that, in the course of this time, a young man fresh from school has to acquaint himself with medicine, surgery, obstetrics, therapeutics, pathology, hygiene, as well as with the anatomy and the physiology of the human body; and that his knowledge should be of such a character that it can be relied upon in any emergency, and always ready for practical application. Consider, in addition, that the medical practitioner may be called upon, at any moment, to give evidence in a court of justice in a criminal case; and that it is therefore well that he should know something of the laws of evidence, and of what we call medical jurisprudence. On a medical certificate, a man may be taken from his home and from his business and confined in a lunatic asylum; surely, therefore, it is desirable that the medical practitioner should have some rational and clear conceptions as to the nature and symptoms of mental disease. Bearing in mind all these requirements of medical education, you will admit that the burden on the young aspirant for the medical profession is somewhat of the heaviest, and that it needs some care to prevent his intellectual back from being broken.

Those who are acquainted with the existing systems of medical education will observe that, long as is the catalogue of studies which I have enumerated, I have omitted to mention several that enter into the usual medical curriculum of the present day. I have said not a word about zoology, comparative anatomy, botany, or materia medica. Assuredly this is from no light estimate of the value or importance of such studies in themselves. It may be taken for granted that I should be the last person in the world to object to the teaching of zoology, or comparative anatomy, in themselves; but I have the strongest feeling that, considering the number and the gravity of those studies through which a medical man must pass, if he is to be competent to discharge the serious duties which devolve upon him, subjects which lie so remote as these do from his practical pursuits should be rigorously excluded. The young man, who has enough to do in order to acquire such familiarity with the structure of the human body as will enable him to perform the operations of surgery, ought not, in my judgment, to be occupied with investigations into the anatomy of crabs and starfishes. Undoubtedly the doctor should know the common poisonous plants of his own country when he sees them; but that knowledge may be obtained by a

few hours devoted to the examination of specimens
of such plants, and the desirableness of such
knowledge is no justification, to my mind, for
spending three months over the study of systematic
botany. Again, materia medica, so far as it is a
knowledge of drugs, is the business of the druggist.
In all other callings the necessity of the division of
labour is fully recognised, and it is absurd to require
of the medical man that he should not avail himself
of the special knowledge of those whose business
it is to deal in the drugs which he uses. It is all
very well that the physician should know that
castor oil comes from a plant, and castoreum from
an animal, and how they are to be prepared; but
for all the practical purposes of his profession that
knowledge is not of one whit more value, has no
more relevancy, than the knowledge of how the
steel of his scalpel is made.

All knowledge is good. It is impossible to say
that any fragment of knowledge, however insignificant or remote from one's ordinary pursuits, may
not some day be turned to account. But in medical
education, above all things, it is to be recollected
that, in order to know a little well, one must be
content to be ignorant of a great deal.

Let it not be supposed that I am proposing to
narrow medical education, or, as the cry is, to lower
the standard of the profession. Depend upon it
there is only one way of really ennobling any calling, and that is to make those who pursue it real

masters of their craft, men who can truly do that which they profess to be able to do, and which they are credited with being able to do by the public. And there is no position so ignoble as that of the so-called " liberally-educated practitioner," who may be able to read Galen in the original; who knows all the plants, from the cedar of Lebanon to the hyssop upon the wall; but who finds himself, with the issues of life and death in his hands, ignorant, blundering, and be-wildered, because of his ignorance of the essential and fundamental truths upon which practice must be based. Moreover, I venture to say, that any man who has seriously studied all the essential branches of medical knowledge; who has the needful acquaintance with the elements of physical science; who has been brought by medical jurisprudence into contact with law; whose study of insanity has taken him into the fields of psychology; has *ipso facto* received a liberal education.

Having lightened the medical curriculum by culling out of it everything which is unessential, we may next consider whether something may not be done to aid the medical student toward the acquirement of real knowledge by modifying the system of examination. In England, within my recollection, it was the practice to require of the medical student attendance on lectures upon the most diverse topics during three years; so that it often happened that he would have to listen, in the course of a day, to four or five lectures upon totally different subjects, in addition to the hours given to dissection and to hospital practice : and he was required to keep all the knowledge he could pick up, in this distracting fashion, at examination point, until, at the end of three years, he was set down to a table and questioned pell-mell upon all the different matters with which he had been striving to make acquaintance. A worse system and one more calculated to obstruct the acquisition of sound knowledge and to give full play to the " crammer " and the " grinder " could hardly have been devised by human ingenuity   Of late years great reforms have taken place. Examinations have been divided so as to diminish the number of subjects among which the attention has to be distributed. Practical examination has been largely introduced ; but there still remains, even under the present system, too much of the old evil inseparable from the contemporaneous pursuit of a multiplicity of diverse studies.

Proposals have recently been made to get rid of general examinations altogether, to permit the student to be examined in each subject at the end of his attendance on the class; and then, in case of the result being satisfactory, to allow him to have done with it ; and I may say that this method has been pursued for many years in the Royal School of Mines in London, and has been found to work very well. It allows the student to concentrate his mind upon what he is about for the time being, and then to dismiss it. Those who are occupied in intellectual work, will, I think, agree with me that it is important, not so much to know a thing, as to have known it, and known it thoroughly. If you have once known a thing in this way it is easy to renew your knowledge when you have forgotten it; and when you begin to take the subject up again, it slides back upon the familiar grooves with great facility.

Lastly comes the question as to how the university may co-operate in advancing medical education. A medical school is strictly a technical school—a school in which a practical profession is taught—while a university ought to be a place in which knowledge is obtained without direct reference to professional purposes. It is clear, therefore, that a university and its antecedent, the school, may best co-operate with the medical school by making due provision for the study of those branches of knowledge which lie at the foundation of medicine.

At present, young men come to the medical schools without a conception of even the elements of physical science; they learn, for the first time, that there are such sciences as physics, chemistry, and physiology, and are introduced to anatomy as a new thing. It may be safely said that, with a large proportion of medical students, much of the

first session is wasted in learning how to learn
—in familiarising themselves with utterly strange
conceptions. and in awakening their dormant and
wholly untrained powers of observation and of
manipulation. It is difficult to over-estimate the
magnitude of the obstacles which are thrown in
the way of scientific training by the existing
system of school education. Not only are men
trained in mere book-work, ignorant of what
observation means, but the habit of learning from
books alone begets a disgust of observation. The
book-learned student will rather trust to what he
sees in a book than to the witness of his own
eyes.

There is not the least reason why this should
be so, and, in fact, when elementary education
becomes that which I have assumed it ought to
be, this state of things will no longer exist.
There is not the slightest difficulty in giving
sound elementary instruction in physics, in
chemistry, and in the elements of human physiology, in ordinary schools. In other words, there
is no reason why the student should not come to
the medical school, provided with as much knowledge of these several sciences as he ordinarily
picks up in the course of his first year of attendance at the medical school.

I am not saying this without full practical
justification for the statement. For the last
eighteen years we have had in England a system

of elementary science teaching carried out under
the auspices of the Science and Art Department,
by which elementary scientific instruction is made
readily accessible to the scholars of all the elementary schools in the country.   Commencing
with small beginnings, carefully developed and
improved, that system now brings up for examination as many as seven thousand scholars in the
subject of human physiology alone.   I can say
that, out of that number, a large proportion have
acquired a fair amount of substantial knowledge;
and that no inconsiderable percentage show as
good an acquaintance with human physiology as
used to be exhibited by the average candidates
for medical degrees in the University of London,
when I was first an examiner there twenty years
ago ; and quite as much knowledge as is possessed
by the ordinary student of medicine at the present
day.   I am justified, therefore, in looking forward
to the time when the student who proposes to
devote himself to medicine will come, not absolutely raw and inexperienced as he is at present,
but in a certain state of preparation for further
study ; and I look to the university to help him
still further forward in that stage of preparation,
through the organisation of its biological department.   Here the student will find means of
acquainting himself with the phenomena of life
in their broadest acceptation.   He will study not
botany and zoology, which, as I have said, would

and give their ability to serve their kind full
play.

I rejoice to observe that the encouragement of
research occupies so prominent a place in your
official documents, and in the wise and liberal
inaugural address of your president. This subject
of the encouragement, or, as it is sometimes called,
the endowment of research, has of late years
greatly exercised the minds of men in England.
It was one of the main topics of discussion by
the members of the Royal Commission of whom
I was one, and who not long since issued their
report, after five years' labour. Many seem to
think that this question is mainly one of money ;
that you can go into the market and buy research,
and that supply will follow demand, as in the
ordinary course of commerce. This view does
not commend itself to my mind. I know of no
more difficult practical problem than the discovery
of a method of encouraging and supporting the
original investigator without opening the door to
nepotism and jobbery. My own conviction is
admirably summed up in the passage of your
president's address, " that the best investigators
are usually those who have also the responsibilities
of instruction, gaining thus the incitement of
colleagues, the encouragement of pupils, and the
observation of the public."

At the commencement of this address I ventured
to assume that I might, if I thought fit, criticise

the arrangements which have been made by the board of trustees, but I confess that I have little to do but to applaud them. Most wise and sagacious seems to me the determination not to build for the present. It has been my fate to see great educational funds fossilise into mere bricks and mortar, in the petrifying springs of architecture, with nothing left to work the institution they were intended to support. A great warrior is said to have made a desert and called it peace. Administrators of educational funds have sometimes made a palace and called it a university. If I may venture to give advice in a matter which lies out of my proper competency, I would say that whenever you do build, get an honest bricklayer, and make him build you just such rooms as you really want, leaving ample space for expansion. And a century hence, when the Baltimore and Ohio shares are at one thousand premium, and you have endowed all the professors you need, and built all the laboratories that are wanted, and have the best museum and the finest library that can be imagined; then, if you have a few hundred thousand dollars you don't know what to do with, send for an architect and tell him to put up a façade. If American is similar to English experience, any other course will probably lead you into having some stately structure, good for your architect's fame, but not in the least what you want.

It appears to me that what I have ventured to lay down as the principles which should govern the relations of a university to education in general, are entirely in accordance with the measures you have adopted. You have set no restrictions upon access to the instruction you propose to give; you have provided that such instruction, either as given by the university or by associated institutions, should cover the field of human intellectual activity. You have recognised the importance of encouraging research. You propose to provide means by which young men, who may be full of zeal for a literary or for a scientific career, but who also may have mistaken aspiration for inspiration, may bring their capacities to a test, and give their powers a fair trial. If such a one fail, his endowment terminates, and there is no harm done. If he succeed, you may give power of flight to the genius of a Davy or a Faraday, a Carlyle or a Locke, whose influence on the future of his fellow-men shall be absolutely incalculable.

You have enunciated the principle that " the glory of the university should rest upon the character of the teachers and scholars, and not upon their numbers or buildings constructed for their use." And I look upon it as an essential and most important feature of your plan that the income of the professors and teachers shall be independent of the number of students whom they can attract. In this way you provide against the danger, patent elsewhere, of finding attempts at improvement obstructed by vested interests; and, in the department of medical education especially, you are free of the temptation to set loose upon the world men utterly incompetent to perform the serious and responsible duties of their profession.

It is a delicate matter for a stranger to the practical working of your institutions, like myself, to pretend to give an opinion as to the organisation of your governing power. I can conceive nothing better than that it should remain as it is, if you can secure a succession of wise, liberal, honest, and conscientious men to fill the vacancies that occur among you. I do not greatly believe in the efficacy of any kind of machinery for securing such a result; but I would venture to suggest that the exclusive adoption of the method of co-optation for filling the vacancies which must occur in your body, appears to me to be somewhat like a tempting of Providence. Doubtless there are grave practical objections to the appointment of persons outside of your body and not directly interested in the welfare of the university; but might it not be well if there were an understanding that your academic staff should be officially represented on the board, perhaps even the heads of one or two independent learned bodies, so that academic opinion and the views of the outside world might potential, wealth in all commodities, and in the energy and ability which turn wealth to account, there is something sublime in the vista of the future. Do not suppose that I am pandering to what is commonly understood by national pride. I cannot say that I am in the slightest degree impressed by your bigness, or your material resources, as such. Size is not grandeur, and territory does not make a nation. The great issue, about which hangs a true sublimity, and the terror of overhanging fate, is what are you going to do with all these things? What is to be the end to which these are to be the means? You are making a novel experiment in politics on the greatest scale which the world has yet seen. Forty millions at your first centenary, it is reasonably to be expected that, at the second, these states will be occupied by two hundred millions of English-speaking people, spread over an area as large as that of Europe, and with climates and interests as diverse as those of Spain and Scandinavia, England and Russia. You and your descendants have to ascertain whether this great mass will hold together under the forms of a republic, and the despotic reality of universal suffrage; whether state rights will hold out against centralisation, without separation; whether centralisation will get the better, without actual or disguised monarchy; whether shifting corruption is better than a permanent bureaucracy; and as population thickens in your

great cities, and the pressure of want is felt, the
gaunt spectre of pauperism will stalk among you,
and communism and socialism will claim to be
heard. Truly America has a great future before
her; great in toil, in care, and in responsibility;
great in true glory if she be guided in wisdom and
righteousness; great in shame if she fail. I cannot
understand why other nations should envy you, or
be blind to the fact that it is for the highest
interest of mankind that you should succeed; but
the one condition of success, your sole safeguard,
is the moral worth and intellectual clearness of the
individual citizen. Education cannot give these,
but it may cherish them and bring them to the
front in whatever station of society they are to be
found; and the universities ought to be, and may
be, the fortresses of the higher life of the nation.

May the university which commences its practical
activity to-morrow abundantly fulfil its high purpose; may its renown as a seat of true learning, a
centre of free inquiry, a focus of intellectual light,
increase year by year, until men wander hither
from all parts of the earth, as of old they sought
Bologna, or Paris, or Oxford.

And it is pleasant to me to fancy that, among
the English students who are drawn to you at that
time, there may linger a dim tradition that a
countryman of theirs was permitted to address you
as he has done to-day, and to feel as if your hopes
were his hopes and your success his joy.

X

ON THE STUDY OF BIOLOGY

[1876]

IT is my duty to-night to speak about the study
of Biology, and while it may be that there are
many of my audience who are quite familiar with
that study, yet as a lecturer of some standing,
it would, I know by experience, be very bad
policy on my part to suppose such to be exten-
sively the case.   On the contrary, I must imagine
that there are many of you who would like to
know what Biology is; that there are others who
have that amount of information, but would nevertheless gladly hear why it should be worth their
while to study Biology; and yet others, again, to
whom these two points are clear, but who desire to
learn how they had best study it, and, finally,
when they had best study it.

I shall, therefore, address myself to the endeavour

to give you some answer to these four questions —what Biology is; why it should be studied; how it should be studied; and when it should be studied.

In the first place, in respect to what Biology is, there are, I believe, some persons who imagine that the term "Biology" is simply a new-fangled denomination, a neologism in short, for what used to be known under the title of "Natural History;" but I shall try to show you, on the contrary, that the word is the expression of the growth of science during the last 200 years, and came into existence half a century ago.

At the revival of learning, knowledge was divided into two kinds—the knowledge of nature and the knowledge of man; for it was the current idea then (and a great deal of that ancient conception still remains) that there was a sort of essential antithesis, not to say antagonism, between nature and man; and that the two had not very much to do with one another, except that the one was oftentimes exceedingly troublesome to the other. Though it is one of the salient merits of our great philosophers of the seventeenth century, that they recognised but one scientific method, applicable alike to man and to nature, we find this notion of the existence of a broad distinction between nature and man in the writings both of Bacon and of Hobbes of Malmesbury; and I have brought with me that famous work which

is now so little known, greatly as it deserves to
be studied, "The Leviathan," in order that I
may put to you in the wonderfully terse and
clear language of Thomas Hobbes, what was
his view of the matter. He says :—

"The register of knowledge of fact is called
history. Whereof there be two sorts, one called
natural history; which is the history of such facts
or effects of nature as have no dependence on
man's will; such as are the histories of metals,
plants, animals, regions, and the like. The other
is civil history; which is the history of the
voluntary actions of men in commonwealths."

So that all history of fact was divided into
these two great groups of natural and of civil history.
The Royal Society was in course of foundation
about the time that Hobbes was writing this
book, which was published in 1651 ; and that
Society was termed a "Society for the Improvement of Natural Knowledge," which was then nearly
the same thing as a " Society for the Improvement of Natural History." As time went on,
and the various branches of human knowledge
became more distinctly developed and separated
from one another, it was found that some were
much more susceptible of precise mathematical
treatment than others. The publication of the
"Principia " of Newton, which probably gave a
greater stimulus to physical science than any work
ever published before, or which is likely to be

published hereafter, showed that precise mathematical methods were applicable to those branches
of science such as astronomy, and what we now
call physics, which occupy a very large portion of
the domain of what the older writers understood
by natural history. And inasmuch as the partly
deductive and partly experimental methods of
treatment to which Newton and others subjected
these branches of human knowledge, showed
that the phenomena of nature which belonged
to them were susceptible of explanation, and
thereby came within the reach of what was called
"philosophy" in those days; so much of this
kind of knowledge as was not included under
astronomy came to be spoken of as "natural philosophy"—a term which Bacon had employed in
a much wider sense. Time went on, and yet
other branches of science developed themselves.
Chemistry took a definite shape; and since all these
sciences, such as astronomy, natural philosophy,
and chemistry, were susceptible either of mathematical treatment or of experimental treatment,
or of both, a broad distinction was drawn between
the experimental branches of what had previously
been called natural history and the observational
branches—those in which experiment was (or
appeared to be) of doubtful use, and where, at
that time, mathematical methods were inapplicable. Under these circumstances the old name
of " Natural History " stuck by the residuum, by

which it did make at the latter end of the last
and the beginning of the present century, thinking men began to discern that under this title
of " Natural History " there were included very
heterogeneous constituents—that, for example,
geology and mineralogy were, in many respects,
widely different from botany and zoology; that a
man might obtain an extensive knowledge of the
structure and functions of plants and animals,
without having need to enter upon the study of
geology or mineralogy, and *vice versa*, and, further
as knowledge advanced, it became clear that there
was a great analogy, a very close alliance, between
those two sciences, of botany and zoology which
deal with human beings, while they are much
more widely separated from all other studies.   It
is due to Buffon to remark that he clearly recognised this great fact. He says : " Ces deux genres
d'êtres organisés [les animaux et les végétaux] ont
beaucoup plus de proprietes communes que de
différences réelles." Therefore, it is not wonderful that, at the beginning of the present century,
in two different countries, and so far as I know,
without any intercommunication, two famous men
clearly conceived the notion of uniting the sciences
which deal with living matter into one whole, and
of dealing with them as one discipline.   In fact,
I may say there were three men to whom this idea
occurred contemporaneously, although there were
but two who carried it into effect, and only one

What ground does it cover ?   I have said that in
its strict technical sense, it denotes all the phenomena which are exhibited by living things, as
distinguished from those which are not living;
but while that is all very well, so long as we
confine ourselves to the lower animals and to
plants, it lands us in considerable difficulties
when we reach the higher forms of living things.
For whatever view we may entertain about the
nature of man, one thing is perfectly certain,
that he is a living creature. Hence, if our definition is to be interpreted strictly, we must include man and all his ways and works under the
head of Biology; in which case, we should find
that psychology, politics, and political economy
would be absorbed into the province of Biology.
In fact, civil history would be merged in natural
history. In strict logic it may be hard to object to
this course, because no one can doubt that the
rudiments and outlines of our own mental phenomena are traceable among the lower animals. They
have their economy and their polity, and if, as is
always admitted, the polity of bees and the
commonwealth of wolves fall within the purview
of the biologist proper, it becomes hard to say why
we should not include therein human affairs,
which, in so many cases, resemble those of the bees
in zealous getting, and are not without a certain
parity in the proceedings of the wolves. The real
fact is that we biologists are a self-sacrificing people;

and inasmuch as, on a moderate estimate, there
are about a quarter of a million different species
of animals and plants to know about already, we
feel that we have more than sufficient territory.
There has been a sort of practical convention by
which we give up to a different branch of science
what Bacon and Hobbes would have called " Civil
History." That branch of science has constituted
itself under the head of Sociology. I may use
phraseology which, at present, will be well understood and say that we have allowed that province
of Biology to become autonomous; but I should
like you to recollect that that is a sacrifice, and
that you should not be surprised if it occasionally
happens that you see a biologist apparently
trespassing in the region of philosophy or politics;
or meddling with human education ; because, after
all, that is a part of his kingdom which he has
only voluntarily forsaken.

Having now defined the meaning of the word
Biology, and having indicated the general scope of
Biological Science, I turn to my second question,
which is—Why should we study Biology ?
Possibly the time may come when that will seem
a very odd question. That we, living creatures,
should not feel a certain amount of interest in
what it is that constitutes our life will eventually,
under altered ideas of the fittest objects of human
inquiry, appear to be a singular phenomenon ; but
at present, judging by the practice of teachers and

educators, Biology would seem to be a topic that does not concern us at all. I propose to put before you a few considerations with which I dare say many will be familiar already, but which will suffice to show—not fully, because to demonstrate this point fully would take a great many lectures —that there are some very good and substantial reasons why it may be advisable that we should know something about this branch of human learning.

I myself entirely agree with another sentiment of the philosopher of Malmesbury, " that the scope of all speculation is the performance of some action or thing to be done," and I have not any very great respect for, or interest in, mere knowing as such. I judge of the value of human pursuits by their bearing upon human interests ; in other words, by their utility ; but I should like that we should quite clearly understand what it is that we mean by this word " utility." In an Englishman's mouth it generally means that by which we get pudding or praise, or both. I have no doubt that is one meaning of the word utility, but it by no means includes all I mean by utility. I think that knowledge of every kind is useful in proportion as it tends to give people right ideas, which are essential to the foundation of right practice, and to remove wrong ideas, which are the no less essential foundations and fertile mothers of every description of error in practice. And inasmuch as, whatever practical people may say, this world is, after all, absolutely governed by ideas, and very often by the wildest and most hypothetical ideas. it is a matter of the very greatest importance that our theories of things, and even of things that seem a long way apart from our daily lives, should be as far as possible true, and as far as possible removed from error. It is not only in the coarser, practical sense of the word " utility," but in this higher and broader sense, that I measure the value of the study of biology by its utility; and I shall try to point out to you that you will feel the need of some knowledge of biology at a great many turns of this present nineteenth century life of ours. For example, most of us attach great importance to the conception which we entertain of the position of man in this universe and his relation to the rest of nature. We have almost all been told, and most of us hold by the tradition, that man occupies an isolated and peculiar position in nature; that though he is in the world he is not of the world; that his relations to things about him are of a remote character; that his origin is recent, his duration likely to be short, and that he is the great central figure round which other things in this world revolve. But this is not what the biologist tells us.

At the present moment you will be kind enough to separate me from them, because it is in no way essential to my present argument that I

should advocate their views. Don't suppose that
I am saying this for the purpose of escaping the
responsibility of their beliefs; indeed, at other
times and in other places, I do not think that
point has been left doubtful; but I want clearly
to point out to you that for my present argument
they may all be wrong; and, nevertheless, my
argument will hold good. The biologists tell us
that all this is an entire mistake. They turn to
the physical organisation of man. They examine
his whole structure, his bony frame and all that
clothes it. They resolve him into the finest particles into which the microscope will enable them
to break him up. They consider the performance
of his various functions and activities, and they
look at the manner in which he occurs on the
surface of the world. Then they turn to other
animals, and taking the first handy domestic
animal—say a dog—they profess to be able to
demonstrate that the analysis of the dog leads
them, in gross, to precisely the same results as the
analysis of the man; that they find almost identi-
cally the same bones, having the same relations;
that they can name the muscles of the dog
by the names of the muscles of the man, and
the nerves of the dog by those of the nerves of
the man, and that, such structures and organs of
sense as we find in the man such also we find in
the dog; they analyse the brain and spinal cord
and they find that the nomenclature which fits,

the one answers for the other. They carry their microscopic inquiries in the case of the dog as far as they can, and they find that his body is resolvable into the same elements as those of the man. Moreover, they trace back the dog's and the man's development, and they find that, at a certain stage of their existence, the two creatures are not distinguishable the one from the other; they find that the dog and his kind have a certain distribution over the surface of the world, comparable in its way to the distribution of the human species. What is true of the dog they tell us is true of all the higher animals; and they assert that they can lay down a common plan for the whole of these creatures, and regard the man and the dog, the horse and the ox as minor modifications of one great fundamental unity. Moreover, the investigations of the last three-quarters of a century have proved, they tell us, that similar inquiries, carried out through all the different kinds of animals which are met with in nature, will lead us, not in one straight series, but by many roads, step by step, gradation by gradation, from man, at the summit, to specks of animated jelly at the bottom of the series. So that the idea of Leibnitz, and of Bonnet, that animals form a great scale of being, in which there are a series of gradations from the most complicated form to the lowest and simplest; that idea, though not exactly in the form in which it was propounded by those philo-

T 2

lower creatures and ourselves, there is one which is hardly ever insisted on,[1] but which may be very fitly spoken of in a place so largely devoted to Art as that in which we are assembled. It is this, that while, among various kinds of animals, it is possible to discover traces of all the other faculties of man, especially the faculty of mimicry, yet that particular form of mimicry which shows itself in the imitation of form, either by modelling or by drawing, is not to be met with. As far as I know, there is no sculpture or modelling, and decidedly no painting or drawing, of animal origin. I mention the fact, in order that such comfort may be derived therefrom as artists may feel inclined to take.

If what the biologists tell us is true, it will be needful to get rid of our erroneous conceptions of man, and of his place in nature, and to substitute right ones for them. But it is impossible to form any judgment as to whether the biologists are right or wrong, unless we are able to appreciate the nature of the arguments which they have to offer.

One would almost think this to be a self-evident proposition. I wonder what a scholar would say to the man who should undertake to criticise a difficult passage in a Greek play, but who obviously had not acquainted himself with

[1] I think that my friend, Professor Allman, was the first to draw attention to it.

the rudiments of Greek grammar. And yet, before
giving positive opinions about these high questions of Biology, people not only do not seem
to think it necessary to be acquainted with
the grammar of the subject, but they have not
even mastered the alphabet. You find criticism
and denunciation showered about by persons who
not only have not attempted to go through the
discipline necessary to enable them to be judges,
but who have not even reached that stage of emer-
gence from ignorance in which the knowledge
that such a discipline is necessary dawns upon the
mind. I have had to watch with some attention—in fact I have been favoured with a good
deal of it myself—the sort of criticism with which
biologists and biological teachings are visited.
I am told every now and then that there is a
" brilliant article "[1] in so-and-so, in which we are
all demolished. I used to read these things once,
but I am getting old now, and I have ceased to
attend very much to this cry of " wolf." When
one does read any of these productions, what one
finds generally, on the face of it is, that the
brilliant critic is devoid of even the elements of
biological knowledge, and that his brilliancy is like

---

[1] Galileo was troubled by a sort of people whom he called
" paper philosophers," because they fancied that the true reading of nature was to be detected by the collation of texts. The
race is not extinct, but, as of old, brings forth its " winds of
doctrine " by which the weathercock heads among us are much
exercised.

the light given out by the crackling of thorns under a pot of which Solomon speaks. So far as I re-collect, Solomon makes use of the image for purposes of comparison; but I will not proceed further into that matter.

Two things must be obvious : in the first place, that every man who has the interests of truth at heart must earnestly desire that every well-founded and just criticism that can be made should be made; but that, in the second place, it is essential to anybody's being able to benefit by criticism, that the critic should know what he is talking about, and be in a position to form a mental image of the facts symbolised by the words he uses. If not, it is as obvious in the case of a biological argument, as it is in that of a his-torical or philological discussion, that such criticism is a mere waste of time on the part of its author, and wholly undeserving of attention on the part of those who are criticised. Take it then as an illustration of the importance of biological study, that thereby alone are men able to form something like a rational conception of what constitutes valuable criticism of the teachings of biologists.[1]

---

[1] Some critics do not even take the trouble to read. I have recently been adjured with much solemnity, to state publicly why I have "changed my opinion" as to the value of the palæontological evidence of the occurrence of evolution.

To this my reply is, Why should I, when that statement was made seven years ago ? An address delivered from the Presi-dential Chair of the Geological Society, in 1870, may be said to

Next, I may mention another bearing of biological knowledge—a more practical one in the
ordinary sense of the word.   Consider the theory
of infectious disease.   Surely that is of interest to
all of us.   Now the theory of infectious disease is
rapidly being elucidated by biological study.   It is
possible to produce, from among the lower animals,
examples of devastating diseases which spread in
the same manner as our infectious disorders, and
which are certainly and unmistakably caused by
living organisms.   This fact renders it possible, at
any rate, that that doctrine of the causation of infectious disease which is known under the name of
" the germ theory " may be well-founded ; and, if
so, it must needs lead to the most important
practical measures in dealing with those terrible
visitations.   It may be well that the general, as
well as the professional, public should have a
sufficient knowledge of biological truths to be able

be a public document, inasmuch as it not only appeared in the
Journal of that learned body, but was re-published, in 1873, in
a volume of *Critiques and Addresses*, to which my name is
attached.   Therein will be found a pretty full statement of my
reasons for enunciating two propositions : (1) that " when we
turn to the higher *Vertebrata*, the results of recent investigations, however we may sift and criticise them, seem to me to
leave a clear balance in favour of the evolution of living forms
one from another ; " and (2) that the case of the horse is one
which " will stand rigorous criticism."
    Thus I do not see clearly in what way I can be said to have
changed my opinion, except in the way of intensifying it, when
in consequence of the accumulation of similar evidence since
1870, I recently spoke of the denial of evolution as not worth
serious consideration.

exactly what the words which he finds in his books and hears from his teachers, mean. If he does not do so, he may read till the crack of doom, but he will never know much about chemistry. That is what every chemist will tell you, and the physicist will do the same for his branch of science. The great changes and improvements in physical and chemical scientific education, which have taken place of late, have all resulted from the combination of practical teaching with the reading of books and with the hearing of lectures. The same thing is true in Biology. Nobody will ever know anything about Biology except in a dilettante "paper-philosopher" way, who contents himself with reading books on botany, zoology, and the like; and the reason of this is simple and easy to understand. It is that all language is merely symbolical of the things of which it treats; the more complicated the things, the more bare is the symbol, and the more its verbal definition requires to be supplemented by the information derived directly from the handling, and the seeing, and the touching of the thing symbolised :—that is really what is at the bottom of the whole matter. It is plain common sense, as all truth, in the long run, is only common sense clarified. If you want a man to be a tea merchant, you don't tell him to read books about China or about tea, but you put him into a tea-merchant's office where he has the handling, the smelling, and the tasting of tea. Without the sort of knowledge which can be gained only in this practical way, his exploits as a tea merchant will soon come to a bankrupt termination. The "paper-philosophers" are under the delusion that physical science can be mastered as literary accomplishments are acquired, but unfortunately it is not so. You may read any quantity of books, and you may be almost as ignorant as you were at starting, if you don't have, at the back of your minds, the change for words in definite images which can only be acquired through the operation of your observing faculties on the phenomena of nature.

It may be said:—"That is all very well, but you told us just now that there are probably something like a quarter of a million different kinds of living and extinct animals and plants, and a human life could not suffice for the examination of one-fiftieth part of all these." That is true, but then comes the great convenience of the way things are arranged ; which is, that although there are these immense numbers of different kinds of living things in existence, yet they are built up, after all, upon marvellously few plans.

There are certainly more than 100,000 species of insects, and yet anybody who knows one insect —if a properly chosen one—will be able to have a very fair conception of the structure of the whole. I do not mean to say he will know that structure thoroughly, or as well as it is desirable he should know it; but he will have enough real knowledge to enable him to understand what he reads, to have genuine images in his mind of those structures which become so variously modified in all the forms of insects he has not seen. In fact, there are such things as types of form among animals and vegetables, and for the purpose of getting a definite knowledge of what constitutes the leading modifications of animal and plant life, it is not needful to examine more than a comparatively small number of animals and plants.

Let me tell you what we do in the biological laboratory which is lodged in a building adjacent to this. There I lecture to a class of students daily for about four-and-a-half months, and my class have, of course, their text-books; but the essential part of the whole teaching, and that which I regard as really the most important part of it, is a laboratory for practical work, which is simply a room with all the appliances needed for ordinary dissection. We have tables properly arranged in regard to light, microscopes, and dissecting instruments, and we work through the structure of a certain number of animals and plants. As, for example, among the plants, we take a yeast plant, a *Protococcus*, a common mould, a *Chara*, a fern, and some flowering plant; among animals we examine such things as an *Amœba*, a *Vorticella*, and a fresh-water polype. We dissect a star-fish, an earth-worm, a snail, a squid, and a fresh-water mussel. We examine a lobster and a cray-fish, and a black beetle. We go on to a common skate, a cod-fish, a frog, a tortoise, a pigeon, and a rabbit, and that takes us about all the time we have to give. The purpose of this course is not to make skilled dissectors, but to give every student a clear and definite conception, by means of sense-images, of the characteristic structure of each of the leading modifications of the animal kingdom ; and that is perfectly possible, by going no further than the length of that list of forms which I have enumerated. If a man knows the structure of the animals I have mentioned, he has a clear and exact, however limited, apprehension of the essential features of the organisation of all those great divisions of the animal and vegetable kingdoms to which the forms I have mentioned severally belong. And it then becomes possible for him to read with profit; because every time he meets with the name of a structure, he has a definite image in his mind of what the name means in the particular creature he is reading about, and therefore the reading is not mere reading. It is not mere repetition of words; but every term employed in the description, we will say, of a horse, or of an elephant, will call up the image of the things he had seen in the rabbit, and he is able to form a distinct conception of that which he has not seen, as a modification of that which he has seen.

That need is not met by constructing a sort of happy hunting-ground of miles of glass cases; and, under the pretence of exhibiting everything putting the maximum amount of obstacle in the way of those who wish properly to see anything.

What the public want is easy and unhindered access to such a collection as they can understand and appreciate; and what the men of science want is similar access to the materials of science. To this end the vast mass of objects of natural history should be divided into two parts—one open to the public, the other to men of science, every day. The former division should exemplify all the more important and interesting forms of life. Explanatory tablets should be attached to them, and catalogues containing clearly-written popular expositions of the general significance of the objects exhibited should be provided. The latter should contain, packed into a comparatively small space, in rooms adapted for working purposes, the objects of purely scientific interest. For example, we will say I am an ornithologist. I go to examine a collection of birds. It is a positive nuisance to have them stuffed. It is not only sheer waste, but I have to reckon with the ideas of the bird-stuffer, while, if I have the skin and nobody has interfered with it, I can form my own judgment as to what the bird was like. For ornithological purposes, what is needed is not glass cases full of stuffed birds on perches, but convenient drawers into each of which a great quantity of skins will go. They occupy no great space and do not require any expenditure beyond their original cost. But for the edification of the public, who want to learn indeed, but do not seek for minute and technical knowledge, the case is different. What one of the general public walking into a collection of birds desires to see is not all the birds that can be got together. He does not want to compare a hundred species of the sparrow tribe side by side; but he wishes to know what a bird is, and what are the great modifications of bird structure, and to be able to get at that knowledge easily. What will best serve his purpose is a comparatively small number of birds carefully selected, and artistically, as well as accurately, set up; with their different ages, their nests, their young, their eggs, and their skeletons side by side; and in accordance with the admirable plan which is pursued in this museum, a tablet, telling the spectator in legible characters what they are and what they mean. For the instruction and recreation of the public such a typical collection would be of far greater value than any many-acred imitation of Noah's ark.

Lastly comes the question as to when biological study may best be pursued. I do not see any valid reason why it should not be made, to a certain extent, a part of ordinary school training.

I have long advocated this view, and I am
perfectly certain that it can be carried out with
ease, and not only with ease, but with very
considerable profit to those who are taught; but
then such instruction must be adapted to the
minds and needs of the scholars. They used to
have a very odd way of teaching the classical
languages when I was a boy. The first task set
you was to learn the rules of the Latin grammar
in the Latin language—that being the language
you were going to learn! I thought then that
this was an odd way of learning a language, but
did not venture to rebel against the judgment of
my superiors. Now, perhaps, I am not so modest
as I was then, and I allow myself to think that it
was a very absurd fashion. But it would be no
less absurd, if we were to set about teaching
Biology by putting into the hands of boys a series
of definitions of the classes and orders of the
animal kingdom, and making them repeat them
by heart. That is so very favourite a method of
teaching, that I sometimes fancy the spirit of the
old classical system has entered into the new
scientific system, in which case I would much
rather that any pretence at scientific teaching
were abolished altogether. What really has to be
done is to get into the young mind some notion
of what animal and vegetable life is. In this
matter, you have to consider practical convenience
as well as other things. There are difficulties in

the way of a lot of boys making messes with
slugs and snails; it might not work in practice.
But there is a very convenient and handy animal
which everybody has at hand, and that is himself;
and it is a very easy and simple matter to obtain
common plants. Hence the general truths of
anatomy and physiology can be taught to young
people in a very real fashion by dealing with the
broad facts of human structure. Such viscera as
they cannot very well examine in themselves,
such as hearts, lungs, and livers, may be obtained
from the nearest butcher's shop. In respect to
teaching something about the biology of plants,
there is no practical difficulty, because almost any
of the common plants will do, and plants do not
make a mess—at least they do not make an
unpleasant mess; so that, in my judgment, the
best form of Biology for teaching to very young
people is elementary human physiology on the
one hand, and the elements of botany on the
other; beyond that I do not think it will be
feasible to advance for some time to come. But
then I see no reason, why, in secondary schools,
and in the Science Classes which are under the
control of the Science and Art Department—
and which I may say, in passing, have in my
judgment, done so very much for the diffusion of
a knowledge of science over the country—we
should not hope to see instruction in the elements
of Biology carried out, not perhaps to the same

recommending him to go through a course of comparative anatomy and physiology, and then to study development. I am sorry to say he was very much displeased, as people often are with good advice. Notwithstanding this discouraging result, I venture, as a parting word, to repeat the suggestion, and to say to all the more or less acute lay and clerical "paper-philosophers"[1] who venture into the regions of biological controversy—Get a little sound, thorough, practical, elementary instruction in biology.

[1] Writers of this stamp are fond of talking about the Baconian method. I beg them therefore to lay to heart these two weighty sayings of the herald of Modern Science :—

"Syllogismus ex propositionibus constat, propositiones ex verbis, verba notionum tesseræ sunt. Itaque si notiones ipsæ (*id quod basis rei est*) confusæ sint et temere a rebus abstractæ, nihil in iis quæ superstruuntur est firmitudinis."—*Novum Organon*, ii. 14.

"Huic autem vanitati nonnulli ex modernis summa levitate ita indulserunt, ut in primo capitulo Geneseos et in libro Job et aliis scripturis sacris, philosophiam naturalem fundare conati sint ; *inter vivos quærentes mortua.*"—*Ibid.* 65.

XI

ON ELEMENTARY INSTRUCTION IN PHYSIOLOGY

[1877]

THE chief ground upon which I venture to recommend that the teaching of elementary physiology should form an essential part of any organised course of instruction in matters pertaining to domestic economy, is, that a knowledge of even the elements of this subject supplies those conceptions of the constitution and mode of action of the living body, and of the nature of health and disease, which prepare the mind to receive instruction from sanitary science.

It is, I think, eminently desirable that the hygienist and the physician should find something in the public mind to which they can appeal; some little stock of universally acknowledged truths, which may serve as a foundation for their warnings, and predispose towards an intelligent obedience to their recommendations.

Listening to ordinary talk about health, disease, and death, one is often led to entertain a doubt whether the speakers believe that the course of natural causation runs as smoothly in the human body as elsewhere. Indications are too often obvious of a strong, though perhaps an unavowed and half unconscious, under-current of opinion that the phenomena of life are not only widely different, in their superficial characters and in their practical importance, from other natural events, but that they do not follow in that definite order which characterises the succession of all other occurrences, and the statement of which we call a law of nature

Hence, I think, arises the want of heartiness of belief in the value of knowledge respecting the laws of health and disease, and of the foresight and care to which knowledge is the essential preliminary, which is so often noticeable ; and a corresponding laxity and carelessness in practice, the results of which are too frequently lamentable.

It is said that among the many religious sects of Russia, there is one which holds that all disease is brought about by the direct and special interference of the Deity, and which, therefore, looks with repugnance upon both preventive and curative measures as alike blasphemous interferences with the will of God. Among ourselves, the " Peculiar People " are, I believe, the only persons who hold the like doctrine in its integrity, and carry it out

with logical rigour. But many of us are old
enough to recollect that the administration of
chloroform in assuagement of the pangs of child-
birth was, at its introduction, strenuously resisted
upon similar grounds.

I am not sure that the feeling, of which the
doctrine to which I have referred is the full
expression, does not lie at the bottom of the
minds of a great many people who yet would
vigorously object to give a verbal assent to the
doctrine itself. However this may be, the main
point is that sufficient knowledge has now been
acquired of vital phenomena, to justify the
assertion, that the notion, that there is anything
exceptional about these phenomena, receives not a
particle of support from any known fact. On the
contrary, there is a vast and an increasing mass of
evidence that birth and death, health and disease,
are as much parts of the ordinary stream of events
as the rising and setting of the sun, or the changes
of the moon ; and that the living body is a
mechanism, the proper working of which we term
health ; its disturbance, disease; its stoppage,
death. The activity of this mechanism is dependent upon many and complicated conditions,
some of which are hopelessly beyond our control,
while others are readily accessible, and are capable
of being indefinitely modified by our own actions.
The business of the hygienist and of the physician
is to know the range of these modifiable conditions,

and how to influence them towards the main-
tenance of health and the prolongation of life;
the business of the general public is to give an
intelligent assent, and a ready obedience based
upon that assent, to the rules laid down for their
guidance by such experts. But an intelligent
assent is an assent based upon knowledge, and the
knowledge which is here in question means an
acquaintance with the elements of physiology.

It is not difficult to acquire such knowledge.
What is true, to a certain extent, of all the physical
sciences, is eminently characteristic of physiology
—the difficulty of the subject begins beyond the
stage of elementary knowledge, and increases with
every stage of progress. While the most highly
trained and the best furnished intellect may find
all its resources insufficient, when it strives to
reach the heights and penetrate into the depths
of the problems of physiology, the elementary
and fundamental truths can be made clear to a
child.

No one can have any difficulty in comprehending the mechanism of circulation or respiration;
or the general mode of operation of the organ of
vision; though the unravelling of all the minutiæ
of these processes, may, for the present, baffle the
conjoined attacks of the most accomplished physicists, chemists, and mathematicians. To know
the anatomy of the human body, with even an
approximation to thoroughness, is the work of a

life; but as much as is needed for a sound comprehension of elementary physiological truths,
may be learned in a week.

A knowledge of the elements of physiology is
not only easy of acquirement, but it may be made
a real and practical acquaintance with the facts,
as far as it goes. The subject of study is always
at hand, in one's self. The principal constituents
of the skeleton, and the changes of form of contracting muscles, may be felt through one's own
skin. The beating of one's heart, and its connection with the pulse, may be noted; the influence
of the valves of one's own veins may be shown;
the movements of respiration may be observed;
while the wonderful phenomena of sensation
afford an endless field for curious and interesting
self-study. The prick of a needle will yield, in a
drop of one's own blood, material for microscopic
observation of phenomena which lie at the foundation of all biological conceptions; and a cold,
with its concomitant coughing and sneezing, may
prove the sweet uses of adversity by helping one
to a clear conception of what is meant by "reflex
action."

Of course there is a limit to this physiological
self-examination. But there is so close a solidar-
ity between ourselves and our poor relations of
the animal world, that our inaccessible inward
parts may be supplemented by theirs. A comparative anatomist knows that a sheep's heart and

lungs, or eye, must not be confounded with those of a man ; but, so far as the comprehension of the elementary facts of the physiology of circulation, of respiration, and of vision goes, the one furnishes the needful anatomical data as well as the other.

Thus, it is quite possible to give instruction in elementary physiology in such a manner as, not only to confer knowledge, which, for the reason I have mentioned, is useful in itself; but to serve the purposes of a training in accurate observation, and in the methods of reasoning of physical science. But that is an advantage which I mention only incidentally, as the present Conference does not deal with education in the ordinary sense of the word.

It will not be suspected that I wish to make physiologists of all the world. It would be as reasonable to accuse an advocate of the "three R's" of a desire to make an orator, an author, and a mathematician of everybody. A stumbling reader, a pot-hook writer, and an arithmetician who has not got beyond the rule of three, is not a person of brilliant acquirements; but the difference between such a member of society and one who can neither read, write, nor cipher is almost inexpressible ; and no one nowadays doubts the value of instruction, even if it goes no farther.

The saying that a little knowledge is a dangerous thing is, to my mind, a very dangerous adage,

If knowledge is real and genuine, I do not believe that it is other than a very valuable possession, however infinitesimal its quantity may be. Indeed, if a little knowledge is dangerous, where is the man who has so much as to be out of danger?

If William Harvey's life-long labours had revealed to him a tenth part of that which may be made sound and real knowledge to our boys and girls, he would not only have been what he was, the greatest physiologist of his age, but he would have loomed upon the seventeenth century as a sort of intellectual portent. Our "little knowledge" would have been to him a great, astounding, unlooked-for vision of scientific truth.

I really see no harm which can come of giving our children a little knowledge of physiology. But then, as I have said, the instruction must be real, based upon observation, eked out by good explanatory diagrams and models, and conveyed by a teacher whose own knowledge has been acquired by a study of the facts; and not the mere catechismal parrot-work which too often usurps the place of elementary teaching.

It is, I hope, unnecessary for me to give a formal contradiction to the silly fiction, which is assiduously circulated by fanatics who not only ought to know, but do know, that their assertions are untrue, that I have advocated the introduction of that experimental discipline which is absolutely indispensable to the professed physiologist, into elementary teaching.

But while I should object to any experimentation which can justly be called painful, for the purpose of elementary instruction; and, while, as a member of a late Royal Commission, I gladly did my best to prevent the infliction of needless pain, for any purpose; I think it is my duty to take this opportunity of expressing my regret at a condition of the law which permits a boy to troll for pike, or set lines with live frog bait, for idle amusement; and, at the same time, lays the teacher of that boy open to the penalty of fine and imprisonment, if he uses the same animal for the purpose of exhibiting one of the most beautiful and instructive of physiological spectacles, the circulation in the web of the foot. No one could undertake to affirm that a frog is not inconvenienced by being wrapped up in a wet rag, and having his toes tied out; and it cannot be denied that inconvenience is a sort of pain. But you must not inflict the least pain on a vertebrated animal for scientific purposes (though you may do a good deal in that way for gain or for sport) without due licence of the Secretary of State for the Home Department, granted under the authority of the Vivisection Act.

So it comes about, that, in this present year of grace 1877, two persons may be charged with cruelty to animals. One has impaled a frog, and suffered the creature to writhe about in that condition for hours; the other has pained the animal no more than one of us would be pained by tying strings round his fingers, and keeping him in the position of a hydropathic patient. The first offender says "I did it because I find fishing very amusing," and the magistrate bids him depart in peace; nay, probably wishes him good sport. The second pleads, "I wanted to impress a scientific truth, with a distinctness attainable in no other way, on the minds of my scholars," and the magistrate fines him five pounds.

I cannot but think that this is an anomalous and not wholly creditable state of things.

you I have done my best to play my part heartily,
and to rejoice in the success of those who have
succeeded. Still, I should like to remind you at
the end of it all, that success on an occasion of
this kind, valuable and important as it is, is in
reality only putting the foot upon one rung of
the ladder which leads upwards; and that the
rung of a ladder was never meant to rest upon,
but only to hold a man's foot long enough to
enable him to put the other somewhat higher.
I trust that you will all regard these successes as
simply reminders that your next business is,
having enjoyed the success of the day, no longer
to look at that success, but to look forward to the
next difficulty that is to be conquered. And now,
having had so much to say to the successful
candidates, you must forgive me if I add that a sort
of under-current of sympathy has been going on in
my mind all the time for those who have not been
successful, for those valiant knights who have
been overthrown in your tourney, and have not
made their appearance in public. I trust that,
in accordance with old custom, they, wounded and
bleeding, have been carried off to their tents, to
be carefully tended by the fairest of maidens;
and in these days, when the chances are that
every one of such maidens will be a qualified
practitioner, I have no doubt that all the splinters
will have been carefully extracted, and that they
are now physically healed. But there may

remain some little fragment of moral or intellectual discouragement, and therefore I will take
the liberty to remark that your chairman to-day,
if he occupied his proper place, would be among
them. Your chairman, in virtue of his position,
and for the brief hour that he occupies that
position, is a person of importance ; and it may
be some consolation to those who have failed if
I say, that the quarter of a century which I have
been speaking of, takes me back to the time when
I was up at the University of London, a candidate
for honours in anatomy and physiology, and when
I was exceedingly well beaten by my excellent
friend, Dr. Ransom, of Nottingham. There is a
person here who recollects that circumstance very
well. I refer to your venerated teacher and mine,
Dr. Sharpey. He was at that time one of the
examiners in anatomy and physiology, and you
may be quite sure that, as he was one of the
examiners, there remained not the smallest doubt
in my mind of the propriety of his judgment, and
I accepted my defeat with the most comfortable
assurance that I had thoroughly well earned it.
But, gentlemen, the competitor having been a
worthy one, and the examination a fair one, I
cannot say that I found in that circumstance
anything very discouraging. I said to myself,
" Never mind ; what's the next thing to be
done ? " And I found that policy of " never

the profession of medicine ; and I do not doubt, from the evidences of ability which have been given to-day, that I have before me a number of men who will rise to eminence in that profession, and who will exert a great and deserved influence upon its future. That in which I am interested, and about which I wish to speak, is the subject of medical education, and I venture to speak about it for the purpose, if I can, of influencing you, who may have the power of influencing the medical education of the future. You may ask, by what authority do I venture, being a person not concerned in the practice of medicine, to meddle with that subject ? I can only tell you it is a fact, of which a number of you I dare say are aware by experience (and I trust the experience has no painful associations), that I have been for a considerable number of years (twelve or thirteen years to the best of my recollection) one of the examiners in the University of London. You are further aware that the men who come up to the University of London are the picked men of the medical schools of London, and therefore such observations as I may have to make upon the state of knowledge of these gentlemen, if they be justified, in regard to any faults I may have to find, cannot be held to indicate defects in the capacity, or in the power of application of those gentlemen, but must be laid, more or less, to the account of the prevalent system of medical educa-

the fact. The knowledge I have looked for was a
real, precise, thorough, and practical knowledge of
fundamentals ; whereas that which the best of the
candidates, in a large proportion of cases, have had
to give me was a large, extensive, and inaccurate
knowledge of superstructure ; and that is what I
mean by saying that my demands went too low
and not too high. What I have had to complain
of is, that a large proportion of the gentlemen
who come up for physiology to the University
of London do not know it as they know their
anatomy, and have not been taught it as they
have been taught their anatomy. Now, I should
not wonder at all if I heard a great many " No,
noes " here ; but I am not talking about University
College ; as I have told you before, I am talking
about the average education of medical schools.
What I have found, and found so much reason to
lament, is, that while anatomy has been taught as
a science ought to be taught, as a matter of
autopsy, and observation, and strict discipline ; in
a very large number of cases, physiology has been
taught as if it were a mere matter of books and of
hearsay. I declare to you, gentlemen, that I have
often expected to be told, when I have asked
a question about the circulation of the blood,
that Professor Breitkopf is of opinion that it
circulates, but that the whole thing is an open
question. I assure you that I am hardly
exaggerating the state of mind on matters of

to study the nature of the works of the human
watch, and the next thing was to study the way
the parts worked together, and the way the watch
worked. Thus, by degrees, we have had growing
up our body of anatomists, or knowers of the construction of the human watch, and our physiologists, who know how the machine works. And
just as any sensible man, who has a valuable
watch, does not meddle with it himself, but goes
to some one who has studied watchmaking, and
understands what the effect of doing this or that
may be; so, I suppose, the man who, having
charge of that valuable machine, his own body,
wants to have it kept in good order, comes to a
professor of the medical art for the purpose of
having it set right, believing that, by deduction
from the facts of structure and from the facts
of function, the physician will divine what may
be the matter with his bodily watch at that
particular time, and what may be the best means
of setting it right. If that may be taken as a
just representation of the relation of the theoretical branches of medicine—what we may call the
institutes of medicine, to use an old term—to the
practical branches, I think it will be obvious to
you that they are of prime and fundamental
importance. Whatever tends to affect the teaching of them injuriously must tend to destroy and
to disorganise the whole fabric of the medical art. I
think every sensible man has seen this long ago;

medical schools, and the number of them, are such as to render it almost impossible that men who confine themselves to the teaching of the theoretical branches of the profession should be able to make their bread by that operation ; and, you know, if a man cannot make his bread he cannot teach—at least his teaching comes to a speedy end. That is a matter of physiology. Anatomy is fairly well taught, because it lies in the direction of practice, and a man is all the better surgeon for being a good anatomist. It does not absolutely interfere with the pursuits of a practical surgeon if he should hold a Chair of Anatomy—though I do not for one moment say that he would not be a better teacher if he did not devote himself to practice. (Applause.) Yes, I know exactly what that cheer means, but I am keeping as carefully as possible from any sort of allusion to Professor Ellis. But the fact is, that even human anatomy has now grown to be so large a matter, that it takes the whole devotion of a man's life to put the great mass of knowledge upon that subject into such a shape that it can be teachable to the mind of the ordinary student. What the student wants in a professor is a man who shall stand between him and the infinite diversity and variety of human knowledge, and who shall gather all that together, and extract from it that which is capable of being assimilated by the mind. That function is a vast and an important one, and unless, in such subjects as anatomy, a man is wholly free from other cares, it is almost impossible that he can perform it thoroughly and well. But if it be hardly possible for a man to pursue anatomy without actually breaking with his profession, how is it possible for him to pursue physiology?

I get every year those very elaborate reports of Henle and Meissner—volumes of, I suppose, 400 pages altogether—and they consist merely of abstracts of the memoirs and works which have been written on Anatomy and Physiology—only abstracts of them! How is a man to keep up his acquaintance with all that is doing in the physiological world—in a world advancing with enormous strides every day and every hour—if he has to be distracted with the cares of practice? You know very well it must be impracticable to do so. Our men of ability join our medical schools with an eye to the future. They take the Chairs of Anatomy or of Physiology; and by and by they leave those Chairs for the more profitable pursuits into which they have drifted by professional success, and so they become clothed, and physiology is bare. The result is, that in those schools in which physiology is thus left to the benevolence, so to speak, of those who have no time to look to it, the effect of such teaching comes out obviously, and is made manifest in what I spoke of just now—the unreality, the bookishness of the knowledge of the taught. And if this is the case in physiology, still more must it be the case in those branches of physics which are the foundation of physiology; although it may be less the case in chemistry, because for an able chemist a certain honourable and independent career lies in the direction of his work, and he is able, like the anatomist, to look upon what he may teach to the student as not absolutely taking him away from his bread-winning pursuits.

But it is of no use to grumble about this state of things unless one is prepared to indicate some sort of practical remedy. And I believe—and I venture to make the statement because I am wholly independent of all sorts of medical schools, and may, therefore, say what I believe without being supposed to be affected by any personal interest—but I say I believe that the remedy for this state of things, for that imperfection of our theoretical knowledge which keeps down the ability of England at the present time in medical matters, is a mere affair of mechanical arrangement; that so long as you have a dozen medical schools scattered about in different parts of the metropolis, and dividing the students among them, so long, in all the smaller schools at any rate, it is impossible that any other state of things than that which I have been depicting should obtain. Professors must live; to live they must occupy themselves with practice, and if they occupy themselves with practice, the pursuit of the abstract branches of science must go to the wall. All this is a plain and obvious matter of common-sense reasoning. I believe you will never alter this state of things until, either by consent or by *force majeure*—and I should be very sorry to see the latter applied—but until there is some new arrangement, and until all the theoretical branches of the profession, the institutes of medicine, are taught in London in not more than one or two, or at the outside three, central institutions, no good will be effected. If that large body of men, the medical students of London, were obliged in the first place to get a knowledge of the theoretical branches of their profession in two or three central schools, there would be abundant means for maintaining able professors—not, indeed, for enriching them, as they would be able to enrich themselves by practice—but for enabling them to make that choice which such men are so willing to make ; namely, the choice between wealth and a modest competency, when that modest competency is to be combined with a scientific career, and the means of advancing knowledge. I do not believe that all the talking about, and tinkering of, medical education will do the slightest good until the fact is clearly recognised, that men must be thoroughly grounded in the theoretical branches of their profession, and that to this end the teaching of those theoretical branches must be confined to two or three centres.

Now let me add one other word, and that is, that if I were a despot, I would cut down these branches to a very considerable extent. The next thing to be done beyond that which I mentioned just now, is to go back to primary education. The great step towards a thorough medical education is to insist upon the teaching of the elements of the physical sciences in all schools, so that medical students shall not go up to the medical colleges utterly ignorant of that with which they have to deal; to insist on the elements of chemistry, the elements of botany, and the elements of physics being taught in our ordinary and common schools, so that there shall be some preparation for the discipline of medical colleges. And, if this reform were once effected, you might confine the " Institutes of Medicine " to physics as applied to physiology—to chemistry as applied to physiology—to physiology itself, and to anatomy. Afterwards, the student, thoroughly grounded in these matters, might go to any hospital he pleased for the purpose of studying the practical branches of his profession. The practical teaching might be made as local as you like ; and you might use to advantage the opportunities afforded by all these local institutions for acquiring a knowledge of the practice of the profession. But you may say:

" This is abolishing a great deal ; you are getting
rid of botany and zoology to begin with." I have
not a doubt that they ought to be got rid of, as
branches of special medical education; they
ought to be put back to an earlier stage, and
made branches of general education. Let me say,
by way of self-denying ordinance, for which you
will, I am sure, give me credit, that I believe that
comparative anatomy ought to be absolutely
abolished. I say so, not without a certain fear of
the Vice-Chancellor of the University of London
who sits upon my left. But I do not think the
charter gives him very much power over me;
moreover, I shall soon come to an end of my
examinership, and therefore I am not afraid, but
shall go on to say what I was going to say, and
that is, that in my belief it is a downright cruelty
—I have no other word for it—to require from
gentlemen who are engaged in medical studies,
the pretence—for it is nothing else, and can be
nothing else, than a pretence—of a knowledge of
comparative anatomy as part of their medical
curriculum. Make it part of their Arts teaching
if you like, make it part of their general education
if you like, make it part of their qualification for
the scientific degree by all means—that is its
proper place; but to require that gentlemen
whose whole faculties should be bent upon the
acquirement of a real knowledge of human physiology should worry themselves with getting up

tions, there would not be ample room for your activity. Let us count up what we have left. I suppose all the time for medical education that can be hoped for is, at the outside, about four years. Well, what have you to master in those four years upon my supposition? Physics applied to physiology; chemistry applied to physiology; physiology; anatomy; surgery; medicine (including therapeutics); obstetrics; hygiene; and medical jurisprudence—nine subjects for four years! And when you consider what those subjects are, and that the acquisition of anything beyond the rudiments of any one of them may tax the energies of a lifetime, I think that even those energies which you young gentlemen have been displaying for the last hour or two might be taxed to keep you thoroughly up to what is wanted for your medical career.

I entertain a very strong conviction that any one who adds to medical education one iota or tittle beyond what is absolutely necessary, is guilty of a very grave offence. Gentlemen, it will depend upon the knowledge that you happen to possess,—upon your means of applying it within your own field of action,—whether the bills of mortality of your district are increased or diminished; and that, gentlemen, is a very serious consideration indeed. And, under those circumstances, the subjects with which you have to deal being so difficult, their extent so enormous, and

XIII

THE STATE AND THE MEDICAL
PROFESSION

[1884]

AT intervals during the last quarter of a century
committees of the Houses of the Legislature and
specially appointed commissions have occupied
themselves with the affairs of the medical profession. Much evidence has been taken, much
wrangling has gone on over the reports of these
bodies; and sometimes much trouble has been
taken to get measures based upon all this work
through Parliament, but very little has been
achieved.

The Bill introduced last session was not more
fortunate than several predecessors. I suppose
that it is not right to rejoice in the misfortunes of
anything, even a Bill; but I confess that this
event afforded me lively satisfaction, for I was a
member of the Royal Commission on the report

of which the Bill was founded, and I did my best to oppose and nullify that report.

That the question must be taken up again and finally dealt with by the Legislature before long cannot be doubted; but in the meanwhile there is time for reflection, and I think that the non-medical public would be wise if they paid a little attention to a subject which is really of considerable importance to them.

The first question which a plain man is disposed to ask himself is, Why should the State interfere with the profession of medicine any more than it does, say, with the profession of engineering? Anybody who pleases may call himself an engineer, and may practice as such. The State confers no title upon engineers, and does not profess to tell the public that one man is a qualified engineer and that another is not so.

The answers which are given to the question are various, and most of them, I think, are bad. A large number of persons seem to be of opinion that the State is bound no less to take care of the general public, than to see that it is protected against incompetent persons, against quacks and medical impostors in general. I do not take that view of the case. I think it is very much wholesomer for the public to take care of itself in this as in all other matters; and although I am not such a fanatic for the liberty of the subject as to plead that interfering with the way in which a

man may choose to be killed is a violation of
that liberty, yet I do think that it is far better to
let everybody do as he likes. Whether that be so
or not, I am perfectly certain that, as a matter of
practice, it is absolutely impossible to prohibit the
practice of medicine by people who have no special
qualification for it. Consider the terrible consequences of attempting to prohibit practice by a
very large class of persons who are certainly not
technically qualified—I am far from saying a word
as to whether they are otherwise qualified or not.
The number of Ladies Bountiful—grandmothers,
aunts, and mothers-in-law—whose chief delight lies
in the administration of their cherished provision
of domestic medicine, is past computation, and
one shudders to think of what might happen if
their energies were turned from this innocuous, if
not beneficent channel, by the strong arm of the
law. But the thing is impracticable.

Another reason for intervention is propounded,
I am sorry to say, by some, though not many,
members of the medical profession, and is simply
an expression of that trades unionism which tends
to infest professions no less than trades.

The general practitioner trying to make both
ends meet on a poor practice, whose medical training has cost him a good deal of time and money,
finds that many potential patients, whose small
fees would be welcome as the little that helps,
prefer to go and get their shilling's worth of

" doctor's stuff" and advice from the chemist and druggist round the corner, who has not paid sixpence for his medical training, because he has never had any.

The general practitioner thinks this is very hard upon him and ought to be stopped. It is perhaps natural that he should think so, though it would be very difficult for him to justify his opinion on any ground of public policy. But the question is really not worth discussion, as it is obvious that it would be utterly impracticable to stop the practice "over the counter" even it it were desirable.

Is a man who has a sudden attack of pain in tooth or stomach not to be permitted to go to the nearest druggist's shop and ask for something that will relieve him? The notion is preposterous. But if this is to be legal, the whole principle of the permissibility of counter practice is granted.

In my judgment the intervention of the State in the affairs of the medical profession can be justified not upon any pretence of protecting the public, and still less upon that of protecting the medical profession, but simply and solely upon the fact that the State employs medical men for certain purposes, and, as employer, has a right to define the conditions on which it will accept service. It is for the interest of the community that no person shall die without there being some official recognition of the cause of his death. It is a matter of the highest importance to the community that, in civil and criminal cases, the law shall be able to have recourse to persons whose evidence may be taken as that of experts; and it will not be doubted that the State has a right to dictate the conditions under which it will appoint persons to the vast number of naval, military, and civil medical offices held directly or indirectly under the Government. Here, and here only, it appears to me, lies the justification for the intervention of the State in medical affairs. It says, or, in my judgment, should say, to the public, "Practice medicine if you like—go to be practised upon by anybody;" and to the medical practitioner, "Have a qualification, or do not have a qualification if people don't mind it ; but if the State is to receive your certificate of death, if the State is to take your evidence as that of an expert, if the State is to give you any kind of civil, or military, or naval appointment, then we can call upon you to comply with our conditions, and to produce evidence that you are, in our sense of the word, qualified. Without that we will not place you in that position." As a matter of fact, that is the relation of the State to the medical profession in this country. For my part, I think it an extremely healthy relation ; and it is one that I should be very sorry to see altered, except in so far that it would certainly be better if greater facilities were given for the swift and sharp punishment of those who pro-

no central authority, there was nothing to prevent any one of those licensing authorities from
granting a licence to any one upon any conditions
it thought fit. The examination might be a sham,
the curriculum might be a sham, the certificate
might be bought and sold like anything in a
shop; or, on the other hand, the examination
might be fairly good and the diploma corre-
spondingly valuable ; but there was not the smallest
guarantee, except the personal character of the
people who composed the administration of each
of these licensing bodies, as to what might happen.
It was possible for a young man to come to
London and to spend two years and six months
of the time of his compulsory three years "walking
the hospitals" in idleness or worse ; he could then,
by putting himself in the hands of a judicious
"grinder" for the remaining six months, pass
triumphantly through the ordeal of one hour's
vivâ voce examination, which was all that was
absolutely necessary, to enable him to be turned
loose upon the public, like death on the pale
horse, "conquering and to conquer," with the full
sanction of the law, as a " qualified practitioner."

It is difficult to imagine, at present, such a
state of things, still more difficult to depict the
consequences of it, because they would appear
like a gross and malignant caricature ; but it may
be said that there was never a system, or want
of system, which was better calculated to ruin

the students who came under it, or to degrade the profession as a whole. My memory goes back to a time when models from whom the Bob Sawyer of the *Pickwick Papers* might have been drawn were anything but rare.

Shortly before my student days, however, the dawn of a better state of things in England began to be visible, in consequence of the establishment of the University of London, and the comparatively very high standard which it placed before its medical graduates.

I say comparatively high standard, for the requirements of the University in those days, and even during the twelve years at a later period, when I was one of the examiners of the medical faculty, were such as would not now be thought more than respectable, and indeed were in many respects very imperfect. But, relatively to the means of learning, the standard was high, and none but the more able and ambitious of the students dreamed of passing the University. Nevertheless, the fact that many men of this stamp did succeed in obtaining their degrees, led others to follow in their steps, and slowly but surely reacted upon the standard of teaching in the better medical schools. Then came the Medical Act of 1858. That Act introduced two immense improvements: one of them was the institution of what is called the Medical Register, upon which the names of all persons recognised by the State as medical practitioners are entered : and the other was the establishment of the Medical Council, which is a kind of Medical Parliament, composed of representatives of the licensing bodies and of leading men in the medical profession nominated by the Crown. The powers given by the Legislature to the Medical Council were found practically to be very limited, but I think that no fair observer of the work will doubt that this much attacked body has excited no small influence in bringing about the great change for the better, which has been effected in the training of men for the medical profession within my recollection.

Another source of improvement must be recognised in the Scottish Universities, and especially in the medical faculty of the University of Edinburgh. The medical education and examinations of this body were for many years the best of their kind in these islands, and I doubt if, at the present moment, the three kingdoms can show a better school of medicine than that of Edinburgh. The vast number of medical students at that University is sufficient evidence of the opinion of those most interested in this subject.

Owing to all these influences, and to the revolution which has taken place in the course of the last twenty years in our conceptions of the proper method of teaching physical science, the training of the medical student in a good school, and the examination test applied by the great majority of the present licensing bodies, reduced now to nineteen, in consequence of the retirement of the Archbishop and the fusion of two of the other licensing bodies, are totally different from what they were even twenty years ago.

I was perfectly astonished, upon one of my sons commencing his medical career the other day, when I contrasted the carefully-watched courses of theoretical and practical instruction, which he is expected to follow with regularity and industry, and the number and nature of the examinations which he will have to pass before he can receive his licence, not only with the monstrous laxity of my own student days, but even with the state of things which obtained when my term of office as examiner in the University of London expired some sixteen years ago.

I have no hesitation in expressing the opinion, which is fully borne out by the evidence taken before the late Royal Commission, that a large proportion of the existing licensing bodies grant their licence on conditions which ensure quite as high a standard as it is practicable or advisable to exact under present circumstances, and that they show every desire to keep pace with the improvements of the times. And I think there can be no doubt that the great majority have so much improved their ways, that their standard is far above that of the ordinary qualification thirty years ago, and I cannot see what excuse there would be for meddling with them if it were not for two other defects which have to be remedied.

Unfortunately there remain two or three black sheep—licensing bodies which simply trade upon their privilege, and sell the cheapest wares they can for shame's sake supply to the bidder. Another defect in the existing system, even where the examination has been so greatly improved as to be good of its kind, is that there are certain licensing bodies which give a qualification for an acquaintance with either medicine or surgery alone, and which more or less ignore obstetrics. This is a revival of the archaic condition of the profession when surgical operations were mostly left to the barbers and obstetrics to the midwives, and when the physicians thought themselves, and were considered by the world, the " superior persons " of the profession. I remember a story was current in my young days of a great court physician who was travelling with a friend, like himself, bound on a visit to a country house. The friend fell down in an apoplectic fit, and the physician refused to bleed him because it was contrary to professional etiquette for a physician to perform that operation. Whether the friend died or whether he got better because he was not bled I do not remember, but the moral of the story is the same. On the other hand, a famous surgeon was asked whether he meant to bring up his son to his own calling, "No," he said, "he is such a fool, I mean to make a physician of him."

Nowadays, it is happily recognised that medicine is one and indivisible, and that no one can properly practice one branch who is not familiar with at any rate the principles of all. Thus the two great things that are wanted now are, in the first place, some means of enforcing such a degree of uniformity upon all the examining bodies that none should present a disgracefully low minimum or pass examination; and the second point is that some body or other shall have the power of enforcing upon every candidate for the licence to practice the study of the three branches, what is called the tripartite qualification. All the members of the late commission were agreed that these were the main points to be attended to in any proposals for the further improvement of medical training and qualification.

But such being the ends in view, our notions as to the best way of attaining them were singularly divergent; so that it came about that eleven commissioners made seven reports. There was one main majority report and six minor reports, which differed more or less from it, chiefly as to the best method of attaining these two objects.

The majority report recommended the adoption of what is known as the conjoint scheme.

According to this plan the power of granting a licence to practise is to be taken away from all the existing bodies, whether they have done well or ill, and to be placed in the hands of a body of delegates (divisional boards), one for each of the three kingdoms. The licence to practise is to be conferred by passing the delegate examination. The licensee may afterwards, if he pleases, go before any of the existing bodies and indulge in the luxury of another examination and the payment of another fee in order to obtain a title, which does not legally place him in any better position than that which he would occupy without it.

Under these circumstances, of course, the only motive for obtaining the degree of a University or the licence of a medical corporation would be the prestige of these bodies. Hence the "black sheep" would certainly be deserted, while those bodies which have acquired a reputation by doing their duty would suffer less.

But, as the majority report proposes that the existing bodies should be compensated for any loss they might suffer out of the fees of the examiners for the State licence, the curious result would be brought about that the profession of the future would be taxed, for all time, for the purpose of handing over to wholly irresponsible bodies a sum, the amount of which would be large for those who had failed in their duty and small for those who had done it.

The scheme in fact involved a perpetual endowment of the "black sheep," calculated on the maximum of their ill-gained profits.[1] I confess that I found myself unable to assent to a plan which, in addition to the rewarding the evil doers, proposed to take away the privileges of a number of examining bodies which confessedly were doing their duty well, for the sake of getting rid of a few who had failed. It was too much like the Chinaman's device of burning down his house to obtain a poor dish of roast pig—uncertain whether in the end he might not find a mere mass of cinders. What we do know is that the great majority of the existing licensing bodies have marvellously improved in the course of the last twenty years, and are improving. What we do not know is that the complicated scheme of the divisional boards will ever be got to work at all.

My own belief is that every necessary reform may be effected, without any interference with vested interests, without any unjust interference with the prestige of institutions which have been,

---

[1] The fees to be paid by candidates for admission to the examinations of the Divisional Board should be of such an amount as will be sufficient to cover the cost of the examinations and the other expenses of the Divisional Board, *and also to provide the sum required to compensate the medical authorities, or such of them as may be entitled to compensation, for any pecuniary losses they may hereafter sustain by reason of the abolition of their privilege of conferring a licence to practise.* Report 50, p. xii.

and still are, extremely valuable, without any question of compensation arising, and by an extremely simple operation. It is only necessary in fact to add a couple of clauses to the Medical Act to this effect: (1) That from and after such a date no person shall be placed upon the Medical Register unless he possesses the threefold qualification. (2) That from and after this date no examination shall be accepted as satisfactory from any licensing body except such as has been carried on in part by examiners appointed by the licensing body, and in part by coadjutor-examiners of equal authority appointed by the Medical Council or other central authority, and acting under their instructions.

In laying down a rule of this kind the State confiscates nothing, and meddles with nobody, but simply acts within its undoubted right of laying down the conditions under which it will confer certain privileges upon medical practitioners. No one can say that the State has not the right to do this; no one can say that the State interferes with any private enterprise or corporate interest unjustly, in laying down its own conditions for its own service. The plan would have the further advantage that all those corporate bodies which have obtained (as many of them have) a great and just prestige by the admirable way in which they have done their work, would reap their just reward in the thronging of students, thenceforward as formerly, to obtain their qualifications; while those who have neglected their duties, who have in some one or two cases, I am sorry to say, absolutely disgraced themselves, would sink into oblivion, and come to a happy and natural euthanasia, in which their misdeeds and themselves would be entirely forgotten.

Two of my colleagues, Professor Turner and Mr. Bryce, M.P., whose practical familiarity with examinations gave their opinions a high value, expressed their substantial approval of this scheme, and I am unable to see the weight of the objections urged against it. It is urged that the difficulty and expense of adequately inspecting so many examinations and of guaranteeing their efficiency would be great, and the difficulty in the way of a fair adjustment of the representation of existing interests and of the representation of new interests upon the general Medical Council would be almost insuperable.

The latter objection is unintelligible to me. I am not aware that any attempt at such adjustment has been fairly discussed, and until that has been done it may be well not to talk about insuperable difficulties. As to the notion that there is any difficulty in getting the coadjutor-examiners, or that the expense will be overwhelming, we have the experience of Scotland, in which every University does, at the present time, appoint its

entirely new world; he addresses himself to a
kind of work of which he has not the smallest
experience. Up to that time his work has been
with books; he rushes suddenly into work with
things, which is as different from work with books
as anything can well be. I am quite sure that a
very considerable number of young men spend a
very large portion of their first session in simply
learning how to learn subjects which are entirely
new to them. And yet recollect that in this
period of four years they have to acquire a
knowledge of all the branches of a great and
responsible practical calling of medicine, surgery,
obstetrics, general pathology, medical jurisprudence, and so forth. Anybody who knows what
these things are, and who knows what is the kind
of work which is necessary to give a man the
confidence which will enable him to stand at the
bedside and say to the satisfaction of his own
conscience what shall be done, and what shall not
be done, must be aware that if a man has only
four years to do all that in he will not have much
time to spare. But that is not all. As I have
said, the young man comes up, probably ignorant
of the existence of science; he has never heard a
word of chemistry, he has never heard a word of
physics, he has not the smallest conception of the
outlines of biological science; and all these things
have to be learned as well and crammed into the
time which in itself is barely sufficient to acquire

a fair amount of that knowledge which is requisite for the satisfactory discharge of his professional duties.

Therefore it is quite clear to me that, somehow or other, the curriculum must be lightened. It is not that any of the subjects which I have mentioned need not to be studied, and may be eliminated. The only alternative therefore is to lengthen the time given to study. Everybody will agree with me that the practical necessities of life in this country are such that, for the average medical practitioner at any rate, it is hopeless to think of extending the period of professional study beyond the age of twenty-two. So that as the period of study cannot be extended forwards, the only thing to be done is to extend it backwards.

The question is how this can be done. My own belief is that if the Medical Council, instead of insisting upon that examination in general education which I am sorry to say I believe to be entirely futile, were to insist upon a knowledge of elementary physics, and chemistry, and biology, they would be taking one of the greatest steps which at present can be made for the improvement of medical education. And the improvement would be this. The great majority of the young men who are going into the profession have practically completed their general education—or they might very well have done so—by the age of sixteen or seventeen. If the interval between this age and that at which they commence their purely medical studies were employed in obtaining a practical acquaintance with elementary physics, chemistry, and biology, in my judgment it would be as good as two years added to the course of medical study. And for two reasons: in the first place, because the subject-matter of that which they would learn is germane to their future studies, and is so much gained; in the second place, because you might clear out of the course of their professional study a great deal which at present occupies time and attention; and last, but not least—probably most—they would then come to their medical studies prepared for that learning from Nature which is what they have to do in the course of becoming skilful medical men, and for which at present they are not in the slightest degree prepared by their previous education.

The second wish I have to express concerns London especially, and I may speak of it briefly as a more economical use of the teaching power in the medical schools. At this present time every great hospital in London—and there are ten or eleven of them—has its complete medical school, in which not only are the branches of practical medicine taught, but also those studies in general science, such as chemistry, elementary physics, general anatomy, and a variety of other topics which are what used to be called (and the term was an extremely useful one) the institutes of medicine. That was all very well half a century ago ; it is all very ill now, simply because those general branches of science, such as anatomy, physiology, chemistry, physiological chemistry, physiological physics, and so forth, have now become so large, and the mode of teaching them is so completely altered, that it is absolutely impossible for any man to be a thoroughly competent teacher of them, or for any student to be effectually taught without the devotion of the whole time of the person who is engaged in teaching. I undertake to say that it is hopelessly impossible for any man at the present time to keep abreast with the progress of physiology unless he gives his whole mind to it; and the bigger the mind is, the more scope he will find for its employment. Again, teaching has become, and must become still more, practical, and that also involves a large expenditure of time. But if a man is to give his whole time to my business he must live by it, and the resources of the schools do not permit them to maintain ten or eleven physiological specialists.

If the students in their first one or two years were taught the institutes of medicine, in two or three central institutions, it would be perfectly easy to have those subjects taught thoroughly and effectually by persons who gave their whole mind and attention to the subject; while at the same time the medical schools at the hospitals would remain what they ought to be—great institutions in which the largest possible opportunities are laid open for acquiring practical acquaintance with the phenomena of disease. So that the preliminary or earlier half of medical education would take place in the central institutions, and the final half would be devoted altogether to practical studies in the hospitals.

I happen to know that this conception has been entertained, not only by myself, but by a great many of those persons who are most interested in the improvement of medical study for a considerable number of years. I do not know whether anything will come of it this half-century or not; but the thing has to be done. It is not a speculative notion; it lies patent to everybody who is accustomed to teaching, and knows what the necessities of teaching are; and I should very much like to see the first step taken—people making up their minds that it has to be done somehow or other.

The last point to which I may advert is one which concerns the action of the profession itself more than anything else. We have arrangements for teaching, we have arrangements for the testing of qualifications, we have marvellous aids and appliances for the treatment of disease in all sorts of ways; but I do not find in London at the present

time, in this little place of four or five million
inhabitants which supports so many things, any
organisation or any arrangement for advancing the
science of medicine, considered as a pure science.
I am quite aware that there are medical societies
of various kinds; I am not ignorant of the lecture-
ships at the College of Physicians and the College
of Surgeons; there is the Brown Institute; and
there is the Society for the Advancement of Medicine by Research, but there is no means, so far as
I know, by which any person who has the inborn
gifts of the investigator and discoverer of new
truth, and who desires to apply that to the
improvement of medical science, can carry out his
intention.   In Paris there is the University of
Paris, which gives degrees; but there are also the
Sorbonne and the Collége de France, places in
which professoriates are established for the express
purpose of enabling men who have the power of
investigation, the power of advancing knowledge
and thereby reacting on practice, to do that which
it is their special mission to do.   I do not know of
anything of the kind in London; and if it should
so happen that a Claude Bernard or a Ludwig
should turn up in London, I really have not the
slightest notion of what we could do with him.
We could not turn him to account, and I think we
should have to export him to Germany or France.
I doubt whether that is a good or a wise condition
of things.   I do not think it is a condition of things

XIV

THE CONNECTION OF THE BIOLOGICAL SCIENCES WITH MEDICINE

[1881]

THE great body of theoretical and practical knowledge which has been accumulated by the labours of some eighty generations, since the dawn of scientific thought in Europe, has no collective English name to which an objection may not be raised; and I use the term "medicine" as that which is least likely to be misunderstood; though, as every one knows, the name is commonly applied, in a narrower sense, to one of the chief divisions of the totality of medical science.

Taken in this broad sense, "medicine" not merely denotes a kind of knowledge, but it comprehends the various applications of that knowledge to the alleviation of the sufferings, the repair of the injuries, and the conservation of the health, of living beings. In fact, the practical aspect of medicine so far dominates over every other, that the "Healing Art" is one of its most widely-received synonyms. It is so difficult to think of medicine otherwise than as something which is necessarily connected with curative treatment, that we are apt to forget that there must be, and is, such a thing as a pure science of medicine—a "pathology" which has no more necessary subservience to practical ends than has zoology or botany.

The logical connection between this purely scientific doctrine of disease, or pathology, and ordinary biology, is easily traced. Living matter is characterised by its innate tendency to exhibit a definite series of the morphological and physiological phenomena which constitute organisation and life. Given a certain range of conditions, and these phenomena remain the same, within narrow limits, for each kind of living thing. They furnish the normal and typical character of the species, and, as such, they are the subject-matter of ordinary biology.

Outside the range of these conditions, the normal course of the cycle of vital phenomena is disturbed ; abnormal structure makes its appearance, or the proper character and mutual adjustment of the functions cease to be preserved. The extent and the importance of these deviations from the typical life may vary indefinitely. They may have no noticeable influence on the general well-being of the economy, or they may favour it. On the other hand, they may be of such a nature as to impede the activities of the organism, or even to involve its destruction.

In the first case, these perturbations are ranged under the wide and somewhat vague category of "variations"; in the second, they are called lesions, states of poisoning, or diseases; and, as morbid states, they lie within the province of pathology. No sharp line of demarcation can be drawn between the two classes of phenomena. No one can say where anatomical variations end and tumours begin, nor where modification of function, which may at first promote health, passes into disease. All that can be said is, that whatever change of structure or function is hurtful belongs to pathology. Hence it is obvious that pathology is a branch of biology; it is the morphology, the physiology, the distribution, the ætiology of abnormal life.

However obvious this conclusion may be now, it was nowise apparent in the infancy of medicine. For it is a peculiarity of the physical sciences that they are independent in proportion as they are imperfect; and it is only as they advance that the bonds which really unite them all become apparent. Astronomy had no manifest connection with terrestrial physics before the publication of the "Principia"; that of chemistry with physics is of still more modern revelation; that of physics and chemistry with physiology, has been stoutly denied within the recollection of most of us, and perhaps still may be.

Or, to take a case which affords a closer parallel with that of medicine. Agriculture has been cultivated from the earliest times, and, from a remote antiquity, men have attained considerable practical skill in the cultivation of the useful plants, and have empirically established many scientific truths concerning the conditions under which they flourish. But, it is within the memory of many of us, that chemistry on the one hand, and vegetable physiology on the other, attained a stage of development such that they were able to furnish a sound basis for scientific agriculture. Similarly, medicine took its rise in the practical needs of mankind. At first, studied without reference to any other branch of knowledge, it long maintained, indeed still to some extent maintains, that independence. Historically, its connection with the biological sciences has been slowly established, and the full extent and intimacy of that connection are only now beginning to be apparent. I trust I have not been mistaken in supposing that an attempt to give a brief sketch of the steps by which a philosophical necessity has become an historical reality, may not be devoid of interest, possibly of instruction, to the members of this great Congress, profoundly interested as all are in the scientific development of medicine.

The history of medicine is more complete and fuller than that of any other science, except, perhaps, astronomy; and, if we follow back the long record as far as clear evidence lights us, we find ourselves taken to the early stages of the civilisation of Greece. The oldest hospitals were the temples of Æsculapius; to these Asclepeia, always erected on healthy sites, hard by fresh springs and surrounded by shady groves, the sick and the maimed resorted to seek the aid of the god of health. Votive tablets or inscriptions recorded the symptoms, no less than the gratitude, of those who were healed ; and, from these primitive clinical records, the half-priestly, half-philosophic caste of the Asclepiads compiled the data upon which the earliest generalisations of medicine, as an inductive science, were based.

In this state, pathology, like all the inductive sciences at their origin, was merely natural history ; it registered the phenomena of disease, classified them, and ventured upon a prognosis, wherever the observation of constant co-existences and sequences suggested a rational expectation of the like recurrence under similar circumstances.

Further than this it hardly went.   In fact, in the then state of knowledge, and in the condition of philosophical speculation at that time, neither the causes of the morbid state, nor the *rationale* of treatment, were likely to be sought for as we seek for them now. The anger of a god was a sufficient reason for the existence of a malady, and a dream ample warranty for therapeutic measures; that a physical phenomenon must needs have a physical cause was not the implied or expressed axiom that it is to us moderns.

The great man whose name is inseparably connected with the foundation of medicine, Hippocrates, certainly knew very little, indeed practically nothing, of anatomy or physiology; and he would, probably, have been perplexed even to imagine the possibility of a connection between the zoological studies of his contemporary Democritus and medicine. Nevertheless, in so far as he, and those who worked before and after him, in the same spirit, ascertained, as matters of experience, that a wound, or a luxation, or a fever, presented such and such symptoms, and that the return of the patient to health was facilitated by such and such measures, they established laws of nature, and began the construction of the science of pathology. All true science begins with empiricism—though all true science is such exactly, in so far as it strives to pass out of the empirical stage into that of the deduction of empirical from more general truths. Thus, it is not wonderful, that the early physicians had little or nothing to do with the development of biological science; and, on the other hand, that the early biologists did not much concern themselves

with medicine.   There is nothing to show that the
Asclepiads took any prominent share in the work
of founding anatomy, physiology, zoology, and
botany.   Rather do these seem to have sprung
from the early philosophers, who were essentially
natural philosophers, animated by the characteristically Greek thirst for knowledge as such.
Pythagoras, Alcmeon, Democritus, Diogenes of
Apollonia, are all credited with anatomical and
physiological investigations; and, though Aristotle
is said to have belonged to an Asclepiad family,
and not improbably owed his taste for anatomical
and zoological inquiries to the teachings of his
father, the physician Nicomachus, the "Historia
Animalium," and the treatise "De Partibus
Animalium," are as free from any allusion to medicine as if they had issued from a modern biological laboratory.

It may be added, that it is not easy to see in
what way it could have benefited a physician of
Alexander's time to know all that Aristotle knew
on these subjects.   His human anatomy was too
rough to avail much in diagnosis; his physiology
was too erroneous to supply data for pathological
reasoning.   But when the Alexandrian school,
with Erasistratus and Herophilus at their head,
turned to account the opportunities of studying
human structure, afforded to them by the
Ptolemies, the value of the large amount of
accurate knowledge thus obtained to the surgeon

for his operations, and to the physician for his diagnosis of internal disorders, became obvious, and a connection was established between anatomy and medicine, which has ever become closer and closer. Since the revival of learning, surgery, medical diagnosis, and anatomy have gone hand in hand. Morgagni called his great work, " De sedibus et causis morborum per anatomen indagatis," and not only showed the way to search out the localities and the causes of disease by anatomy, but himself travelled wonderfully far upon the road. Bichat, discriminating the grosser constituents of the organs and parts of the body, one from another, pointed out the direction which modern research must take; until, at length, histology, a science of yesterday, as it seems to many of us, has carried the work of Morgagni as far as the microscope can take us, and has extended the realm of pathological anatomy to the limits of the invisible world.

Thanks to the intimate alliance of morphology with medicine, the natural history of disease has, at the present day, attained a high degree of perfection. Accurate regional anatomy has rendered practicable the exploration of the most hidden parts of the organism, and the determination, during life, of morbid changes in them; anatomical and histological post-mortem investigations have supplied physicians with a clear basis upon which to rest the classification of diseases, and with unerring tests of the accuracy or inaccuracy of their diagnoses.

If men could be satisfied with pure knowledge, the extreme precision with which, in these days, a sufferer may be told what is happening, and what is likely to happen, even in the most recondite parts of his bodily frame, should be as satisfactory to the patient as it is to the scientific pathologist who gives him the information. But I am afraid it is not; and even the practising physician, while nowise under-estimating the regulative value of accurate diagnosis, must often lament that so much of his knowledge rather prevents him from doing wrong than helps him to do right.

A scorner of physic once said that nature and disease may be compared to two men fighting, the doctor to a blind man with a club, who strikes into the *mêlée*, sometimes hitting the disease, and sometimes hitting nature. The matter is not mended if you suppose the blind man's hearing to be so acute that he can register every stage of the struggle, and pretty clearly predict how it will end. He had better not meddle at all, until his eyes are opened, until he can see the exact position of the antagonists, and make sure of the effect of his blows. But that which it behoves the physician to see, not, indeed, with his bodily eye, but with clear, intellectual vision, is a process, and the chain of causation involved in that process. Disease, as we

A A 2

have seen, is a perturbation of the normal activities
of a living body, and it is, and must remain, unintelligible, so long as we are ignorant of the nature
of these normal activities. In other words, there
could be no real science of pathology until the
science of physiology had reached a degree of
perfection unattained, and indeed unattainable,
until quite recent times.

So far as medicine is concerned, I am not sure
that physiology, such as it was down to the time
of Harvey, might as well not have existed. Nay,
it is perhaps no exaggeration to say that, within
the memory of living men, justly renowned
practitioners of medicine and surgery knew less
physiology than is now to be learned from the
most elementary text-book; and, beyond a few
broad facts, regarded what they did know as of
extremely little practical importance. Nor am I
disposed to blame them for this conclusion;
physiology must be useless, or worse than useless,
to pathology, so long as its fundamental conceptions are erroneous.

Harvey is often said to be the founder of
modern physiology; and there can be no question
that the elucidations of the function of the
heart, of the nature of the pulse, and of the course
of the blood, put forth in the ever-memorable
little essay, "De motu cordis," directly worked a
revolution in men's views of the nature and of the
concatenation of some of the most important

physiological processes among the higher animals; while, indirectly, their influence was perhaps even more remarkable.

But, though Harvey made this signal and perennially important contribution to the physiology of the moderns, his general conception of vital processes was essentially identical with that of the ancients; and, in the "Exercitationes de generatione," and notably in the singular chapter "De calido innato," he shows himself a true son of Galen and of Aristotle.

For Harvey, the blood possesses powers superior to those of the elements; it is the seat of a soul which is not only vegetative, but also sensitive and motor. The blood maintains and fashions all parts of the body, " idque summâ cum providentiâ et intellectu in finem certum agens, quasi ratiocinio quodam uteretur."

Here is the doctrine of the " pneuma," the product of the philosophical mould into which the animism of primitive men ran in Greece, in full force. Nor did its strength abate for long after Harvey's time. The same ingrained tendency of the human mind to suppose that a process is explained when it is ascribed to a power of which nothing is known except that it is the hypothetical agent of the process, gave rise, in the next century, to the animism of Stahl; and, later, to the doctrine of a vital principle, that " asylum ignorantiæ of physiologists, which has so easily accounted for

everything and explained nothing, down to our
own times.

Now the essence of modern, as contrasted with
ancient, physiological science appears to me to
lie in its antagonism to animistic hypotheses and
animistic phraseology. It offers physical explanations of vital phenomena, or frankly confesses that
it has none to offer. And, so far as I know, the
first person who gave expression to this modern
view of physiology, who was bold enough to
enunciate the proposition that vital phenomena,
like all the other phenomena of the physical
world, are, in ultimate analysis, resolvable into
matter and motion, was René Descartes.

The fifty-four years of life of this most original
and powerful thinker are widely overlapped, on
both sides, by the eighty of Harvey, who survived
his younger contemporary by seven years, and takes
pleasure in acknowledging the French philosopher's appreciation of his great discovery.

In fact, Descartes accepted the doctrine of the
circulation as propounded by " Harvæus médecin
d'Angleterre," and gave a full account of it in his
first work, the famous " Discours de la Méthode,"
which was published in 1637, only nine years
after the exercitation " De motu cordis " ; and,
though differing from Harvey on some important
points (in which it may be noted, in passing,
Descartes was wrong and Harvey right), he always
speaks of him with great respect. And so impor-

tant does the subject seem to Descartes, that he returns to it in the " Traité des Passions," and in the " Traité de l'Homme."

It is easy to see that Harvey's work must have had a peculiar significance for the subtle thinker, to whom we owe both the spiritualistic and the materialistic philosophies of modern times. It was in the very year of its publication, 1628, that Descartes withdrew into that life of solitary investigation and meditation of which his philosophy was the fruit. And, as the course of his speculations led him to establish an absolute distinction of nature between the material and the mental worlds, he was logically compelled to seek for the explanation of the phenomena of the material world within itself; and having allotted the realm of thought to the soul, to see nothing but extension and motion in the rest of nature. Descartes uses " thought " as the equivalent of our modern term " consciousness." Thought is the function of the soul, and its only function. Our natural heat and all the movements of the body, says he, do not depend on the soul. Death does not take place from any fault of the soul, but only because some of the principal parts of the body become corrupted. The body of a living man differs· from that of a dead man in the same way as a watch or other automaton (that is to say, a machine which moves of itself) when it is wound up and has, in itself, the physical principle of the movements which the mechanism is adapted to perform, differs from the same watch, or other machine, when it is broken, and the physical principle of its movement no longer exists. All the actions which are common to us and the lower animals depend only on the conformation of our organs, and the course which the animal spirits take in the brain, the nerves, and the muscles; in the same way as the movement of a watch is produced by nothing but the force of its spring and the figure of its wheels and other parts.

Descartes' "Treatise on Man" is a sketch of human physiology, in which a bold attempt is made to explain all the phenomena of life, except those of consciousness, by physical reasonings. To a mind turned in this direction, Harvey's exposition of the heart and vessels as a hydraulic mechanism must have been supremely welcome.

Descartes was not a mere philosophical theorist, but a hardworking dissector and experimenter, and he held the strongest opinion respecting the practical value of the new conception which he was introducing. He speaks of the importance of preserving health, and of the dependence of the mind on the body being so close that, perhaps, the only way of making men wiser and better than they are, is to be sought in medical science. "It is true," says he, " that as medicine is now practised it contains little that is very useful; but without any desire to depreciate, I am sure that there is no one, even among professional men, who will not declare that all we know is very little as compared with that which remains to be known; and that we might escape an infinity of diseases of the mind, no less than of the body, and even perhaps from the weakness of old age, if we had sufficient knowledge of their causes, and of all the remedies with which nature has provided us."[1]   So strongly impressed was Descartes with this, that he resolved to spend the rest of his life in trying to acquire such a knowledge of nature as would lead to the construction of a better medical doctrine.[2]   The anti-Cartesians found material for cheap ridicule in these aspirations of the philosopher; and it is almost needless to say that, in the thirteen years which elapsed between the publication of the "Discours" and the death of Descartes, he did not contribute much to their realisation.   But, for the next century, all progress in physiology took place along the lines which Descartes laid down.

The greatest physiological and pathological work of the seventeenth century, Borelli's treatise " De Motu Animalium," is, to all intents and purposes, a development of Descartes' fundamental conception ; and the same may be said of the physiology and pathology of   Boerhaave, whose authority dominated in the medical world of the first half of the eighteenth century.

[1] *Discours de la Méthode*, 6e partie, Ed. Cousin, p. 193.
[2] *Ibid.* pp. 193 and 211.

With the origin of modern chemistry, and of electrical science, in the latter half of the eighteenth century, aids in the analysis of the phenomena of life, of which Descartes could not have dreamed, were offered to the physiologist. And the greater part of the gigantic progress which has been made in the present century is a justification of the prevision of Descartes. For it consists, essentially, in a more and more complete resolution of the grosser organs of the living body into physico-chemical mechanisms.

" I shall try to explain our whole bodily machinery in such a way, that it will be no more necessary for us to suppose that the soul produces such movements as are not voluntary, than it is to think that there is in a clock a soul which causes it to show the hours." [1]   These words of Descartes might be appropriately taken as a motto by the author of any modern treatise on physiology.

But though, as I think, there is no doubt that Descartes was the first to propound the fundamental conception of the living body as a physical mechanism, which is the distinctive feature of modern, as contrasted with ancient physiology, he was misled by the natural temptation to carry out, in all its details, a parallel between the machines with which he was familiar, such as clocks and pieces of hydraulic apparatus, and the living machine. In all such machines there is a

[1] *De la Formation du Fœtus.*

central source of power, and the parts of the machine are merely passive distributors of that power. The Cartesian school conceived of the living body as a machine of this kind; and herein they might have learned from Galen, who, whatever ill use he may have made of the doctrine of "natural faculties," nevertheless had the great merit of perceiving that local forces play a great part in physiology.

The same truth was recognised by Glisson, but it was first prominently brought forward in the Hallerian doctrine of the "vis insita" of muscles. If muscle can contract without nerve, there is an end of the Cartesian mechanical explanation of its contraction by the influx of animal spirits.

The discoveries of Trembley tended in the same direction. In the freshwater *Hydra*, no trace was to be found of that complicated machinery upon which the performance of the functions in the higher animals was supposed to depend. And yet the hydra moved, fed, grew, multiplied, and its fragments exhibited all the powers of the whole. And, finally, the work of Caspar F. Wolff,[1] by demonstrating the fact that the growth and development of both plants and animals take place antecedently to the existence of their grosser organs, and are, in fact, the causes and not the consequences of organisation (as then understood), sapped the foundations of the

[1] *Theoria Generationis,* 1759.

Cartesian physiology as a complete expression of vital phenomena.

For Wolff, the physical basis of life is a fluid, possessed of a "vis essentialis" and a "solidescibilitas," in virtue of which it gives rise to organisation; and, as he points out, this conclusion strikes at the root of the whole iatro-mechanical system.

In this country, the great authority of John Hunter exerted a similar influence; though it must be admitted that the too sibylline utterances which are the outcome of Hunter's struggles to define his conceptions are often susceptible of more than one interpretation. Nevertheless, on some points Hunter is clear enough. For example, he is of opinion that "Spirit is only a property of matter" ("Introduction to Natural History," p. 6), he is prepared to renounce animism, (*l.c.* p. 8), and his conception of life is so completely physical that he thinks of it as something which can exist in a state of combination in the food. "The aliment we take in has in it, in a fixed state, the real life; and this does not become active until it has got into the lungs; for there it is freed from its prison" ("Observations on Physiology," p. 113). He also thinks that "It is more in accord with the general principles of the animal machine to suppose that none of its effects are produced from any mechanical principle whatever; and that every effect is produced from

an action in the part; which action is produced
by a stimulus upon the part which acts, or upon
some other part with which this part sympathises
so as to take up the whole action" (*l.c.* p. 152).

And Hunter is as clear as Wolff, with whose
work he was probably unacquainted, that "whatever life is, it most certainly does not depend
upon structure or organisation" (*l.c.* p. 114).

Of course it is impossible that Hunter could
have intended to deny the existence of purely
mechanical operations in the animal body. But
while, with Borelli and Boerhaave, he looked
upon absorption, nutrition, and secretion as
operations effected by means of the small vessels,
he differed from the mechanical physiologists,
who regarded these operations as the result of
the mechanical properties of the small vessels,
such as the size, form, and disposition of their
canals and apertures. Hunter, on the contrary,
considers them to be the effect of properties of
these vessels which are not mechanical but vital.
" The vessels," says he, " have more of the polypus
in them than any other part of the body," and he
talks of the " living and sensitive principles of the
arteries," and even of the " dispositions or feelings
of the arteries." " When the blood is good and
genuine the sensations of the arteries, or the
dispositions for sensation, are agreeable. . . . It
is then they dispose of the blood to the best
advantage, increasing the growth of the whole,

supplying any losses, keeping up a due succession, etc." (*l.c.* p. 133).

If we follow Hunter's conceptions to their logical issue, the life of one of the higher animals is essentially the sum of the lives of all the vessels, each of which is a sort of physiological unit, answering to a polype ; and, as health is the result of the normal "action of the vessels," so is disease an effect of their abnormal action. Hunter thus stands in thought, as in time, midway between Borelli on the one hand, and Bichat on the other.

The acute founder of general anatomy, in fact, outdoes Hunter in his desire to exclude physical reasonings from the realm of life. Except in the interpretation of the action of the sense organs, he will not allow physics to have anything to do with physiology.

" To apply the physical sciences to physiology is to explain the phenomena of living bodies by the laws of inert bodies. Now this is a false principle, hence all its consequences are marked with the same stamp. Let us leave to chemistry its affinity ; to physics, its elasticity and its gravity. Let us invoke for physiology only sensibility and contractility." [1]

Of all the unfortunate dicta of men of eminent ability this seems one of the most unhappy, when we think of what the application of the methods and the data of physics and chemistry has done

[1] *Anatomie générale,* i. p. liv.

towards bringing physiology into its present state. It is not too much to say that one-half of a modern text-book of physiology consists of applied physics and chemistry; and that it is exactly in the exploration of the phenomena of sensibility and contractility that physics and chemistry have exerted the most potent influence.

Nevertheless, Bichat rendered a solid service to physiological progress by insisting upon the fact that what we call life, in one of the higher animals, is not an indivisible unitary archæus dominating, from its central seat, the parts of the organism, but a compound result of the synthesis of the separate lives of those parts.

"All animals," says he, "are assemblages of different organs, each of which performs its function and concurs, after its fashion, in the preservation of the whole. They are so many special machines in the general machine which constitutes the individual. But each of these special machines is itself compounded of many tissues of very different natures, which in truth constitute the elements of those organs " (*l.c.* lxxix.). "The conception of a proper vitality is applicable only to these simple tissues, and not to the organs themselves " (*l.c.* lxxxiv.).

And Bichat proceeds to make the obvious application of this doctrine of synthetic life, if I may so call it, to pathology. Since diseases are only alterations of vital properties, and the properties of each tissue are distinct from those of the rest, it is evident that the diseases of each tissue must be different from those of the rest. Therefore, in any organ composed of different tissues, one may be diseased and the other remain healthy; and this is what happens in most cases (*l.c.* lxxxv.).

In a spirit of true prophecy, Bichat says, " We have arrived at an epoch in which pathological anatomy should start afresh." For, as the analysis of the organs had led him to the tissues as the physiological units of the organism; so, in a succeeding generation, the analysis of the tissues led to the cell as the physiological element of the tissues. The contemporaneous study of development brought out the same result; and the zoologists and botanists, exploring the simplest and the lowest forms of animated beings, confirmed the great induction of the cell theory. Thus the apparently opposed views, which have been battling with one another ever since the middle of the last century, have proved to be each half the truth.

The proposition of Descartes that the body of a living man is a machine, the actions of which are explicable by the known laws of matter and motion, is unquestionably largely true. But it is also true, that the living body is a synthesis of innumerable physiological elements, each of which may nearly be described, in Wolff's words, as a fluid possessed of a "vis essentialis" and a "solidescibilitas"; or, in modern phrase, as protoplasm susceptible of structural mètamorphosis and functional metabolism: and that the only machinery, in the precise sense in which the Cartesian school understood mechanism, is, that which co-ordinates and regulates these physiological units into an organic whole.

In fact, the body is a machine of the nature of an army, not of that of a watch or of a hydraulic apparatus. Of this army each cell is a soldier, an organ a brigade, the central nervous system headquarters and field telegraph, the alimentary and circulatory system the commissariat. Losses are made good by recruits born in camp, and the life of the individual is a campaign, conducted successfully for a number of years, but with certain defeat in the long run.

The efficacy of an army, at any given moment, depends on the health of the individual soldier, and on the perfection of the machinery by which he is led and brought into action at the proper time; and, therefore, if the analogy holds good, there can be only two kinds of diseases, the one dependent on abnormal states of the physiological units, the other on perturbations of their co-ordinating and alimentative machinery.

Hence, the establishment of the cell theory, in normal biology, was swiftly followed by a "cellular pathology," as its logical counterpart. I need not remind you how great an instrument of investigation this doctrine has proved in the hands of the man of genius to whom its development is due, and who would probably be the last to forget that abnormal conditions of the co-ordinative and distributive machinery of the body are no less important factors of disease.

Henceforward, as it appears to me, the connection of medicine with the biological sciences is clearly indicated. Pure pathology is that branch of biology which defines the particular perturbation of cell-life, or of the co-ordinating machinery, or of both, on which the phenomena of disease depend.

Those who are conversant with the present state of biology will hardly hesitate to admit that the conception of the life of one of the higher animals as the summation of the lives of a cell aggregate, brought into harmonious action by a co-ordinative machinery formed by some of these cells, constitutes a permanent acquisition of physiological science. But the last form of the battle between the animistic and the physical views of life is seen in the contention whether the physical analysis of vital phenomena can be carried beyond this point or not.

There are some to whom living protoplasm is a substance, even such as Harvey conceived the blood to be, " summâ cum providentiâ et intellectu in finem certum agens, quasi ratiocinio quodam ; " and who look with as little favour as Bichat did, upon any attempt to apply the principles and the methods of physics and chemistry to the investigation of the vital processes of growth, metabolism, and contractility. They stand upon the ancient ways; only, in accordance with that progress towards democracy, which a great political writer has declared to be the fatal characteristic of modern times, they substitute a republic formed by a few billion of "animulæ" for the monarchy of the all-pervading "anima."

Others, on the contrary, supported by a robust faith in the universal applicability of the principles laid down by Descartes, and seeing that the actions called "vital" are, so far as we have any means of knowing, nothing but changes of place of particles of matter, look to molecular physics to achieve the analysis of the living protoplasm itself into a molecular mechanism. If there is any truth in the received doctrines of physics, that contrast between living and inert matter, on which Bichat lays so much stress, does not exist. In nature, nothing is at rest, nothing is amorphous; the simplest particle of that which men in their blindness are pleased to call "brute matter" is a vast aggregate of molecular mechanisms performing complicated movements of immense rapidity, and sensitively adjusting themselves to every change in the surrounding world. Living matter differs from other matter in degree and not in kind; the microcosm repeats the macrocosm; and one chain of causation connects the nebulous original of suns and planetary systems with the protoplasmic foundation of life and organisation.

From this point of view, pathology is the analogue of the theory of perturbations in astronomy; and therapeutics resolves itself into the discovery of the means by which a system of forces competent to eliminate any given perturbation may be introduced into the economy. And, as pathology bases itself upon normal physiology, so therapeutics rests upon pharmacology; which is, strictly speaking, a part of the great biological topic of the influence of conditions on the living organism, and has no scientific foundation apart from physiology.

It appears to me that there is no more hopeful indication of the progress of medicine towards the ideal of Descartes than is to be derived from a comparison of the state of pharmacology, at the present day, with that which existed forty years ago. If we consider the knowledge positively acquired, in this short time, of the *modus operandi* of urari, of atropia, of physostigmin, of veratria, of casca, of strychnia, of bromide of potassium, of phosphorus, there can surely be no ground for doubting that, sooner or later, the pharmacologist will supply the physician with the means of affecting, in any desired sense, the functions of any physiological element of the body. It will, in short, become possible to introduce into the economy a molecular mechanism which, like a very cunningly-contrived torpedo, shall find its way to some particular group of living elements, and cause an explosion among them, leaving the rest untouched.

The search for the explanation of diseased states in modified cell-life; the discovery of the important part played by parasitic organisms in the ætiology of disease; the elucidation of the action of medicaments by the methods and the data of experimental physiology; appear to me to be the greatest steps which have ever been made towards the establishment of medicine on a scientific basis. I need hardly say they could not have been made except for the advance of normal biology.

There can be no question, then, as to the nature or the value of the connection between medicine and the biological sciences. There can be no doubt that the future of pathology and of therapeutics, and, therefore, that of practical medicine, depends upon the extent to which those who occupy themselves with these subjects are trained in the methods and impregnated with the fundamental truths of biology.

And, in conclusion, I venture to suggest that the collective sagacity of this congress could occupy itself with no more important question than with this: How is medical education to be arranged, so that, without entangling the student in those details of the systematist which are valueless to him, he may be enabled to obtain a firm grasp of the great truths respecting animal and vegetable life, without which, notwithstanding all the progress of scientific medicine, he will still find himself an empiric?

XV

THE SCHOOL BOARDS : WHAT THEY
CAN DO, AND WHAT THEY MAY DO.

[1870]

AN electioneering manifesto would be out of place
in the pages of this Review ; but any suspicion that
may arise in the mind of the reader that the
following pages partake of that nature, will be dispelled, if he reflect that they cannot be published [1]
until after the day on which the ratepayers of the
metropolis will have decided which candidates for
seats upon the Metropolitan School Board they
will take, and which they will leave.

As one of those candidates, I may be permitted
to say, that I feel much in the frame of mind of
the Irish bricklayer's labourer, who bet another

---

[1] Notwithstanding Mr. Huxley's intentions, the Editor
took upon himself, in what seemed to him to be the public
interest, to send an extract from this article to the newspapers
—before the day of the election of the School Board.—EDITOR
of the *Contemporary Review.*

that he could not carry him to the top of the ladder in his hod. The challenged hodman won his wager, but as the stakes were handed over, the challenger wistfully remarked, "I'd great hopes of falling at the third round from the top." And, in view of the work and the worry which awaits the members of the School Boards, I must confess to an occasional ungrateful hope that the friends who are toiling upwards with me in their hod, may, when they reach "the third round from the top," let me fall back into peace and quietness.

But whether fortune befriend me in this rough method, or not, I should like to submit to those of whom I am potential, but of whom I may not be an actual, colleague, and to others who may be interested in this most important problem— how to get the Education Act to work efficiently —some considerations as to what are the duties of the members of the School Boards, and what are the limits of their power.

I suppose no one will be disposed to dispute the proposition, that the prime duty of every member of such a Board is to endeavour to administer the Act honestly ; or in accordance, not only with its letter, but with its spirit. And if so, it would seem that the first step towards this very desirable end is, to obtain a clear notion of what that letter signifies, and what that spirit implies ; or, in other words, what the clauses of the Act are intended to enjoin and to forbid. So that it is really not

admissible, except for factious and abusive purposes, to assume that any one who endeavours to
get at this clear meaning is desirous only of raising
quibbles and making difficulties.

Reading the Act with this desire to understand
it, I find that its provisions may be classified, as
might naturally be expected, under two heads :
the one set relating to the subject-matter of
education ; the other to the establishment, main-
tenance, and administration of the schools in
which that education is to be conducted.

Now it is a most important circumstance, that
all the sections of the Act, except four, belong to
the latter division ; that is, they refer to mere
matters of administration. The four sections in
question are the seventh, the fourteenth, the
sixteenth, and the ninety-seventh. Of these, the
seventh, the fourteenth, and the ninety-seventh
deal with the subject-matter of education, while
the sixteenth defines the nature of the relations
which are to exist between the "Education
Department" (an euphemism for the future
Minister of Education) and the School Boards.
It is the sixteenth clause which is the most
important, and, in some respects, the most remarkable of all. It runs thus :—

"If the School Board do, or permit, any act in contravention of, or fail to comply with, the regulations, according to
which a school provided by them is required by this Act to be
conducted, the Education Department may declare the School

Board to be, and such Board shall accordingly be deemed to be, a Board in default, and the Education Department may proceed accordingly ; and every act, or omission, of any member of the School Board, or manager appointed by them, or any person under the control of the Board, shall be deemed to be permitted by the Board, unless the contrary be proved.

" If any dispute arises as to whether the School Board have done, or permitted, any act in contravention of, or have failed to comply with, the said regulations, *the matter shall be referred to the Education Department, whose decision thereon shall be final.*"

It will be observed that this clause gives the Minister of Education absolute power over the doings of the School Boards. He is not only the administrator of the Act, but he is its interpreter. I had imagined that on the occurrence of a dispute, not as regards a question of pure administration, but as to the meaning of a clause of the Act, a case might be taken and referred to a court of justice. But I am led to believe that the Legislature has, in the present instance, deliberately taken this power out of the hands of the judges and lodged it in those of the Minister of Education, who, in accordance with our method of making Ministers, will necessarily be a political partisan, and who may be a strong theological sectary into the bargain. And I am informed by members of Parliament who watched the progress of the Act, that the responsibility for this unusual state of things rests, not with the Government, but with the Legislature, which exhibited a singular disposition to accumulate power in the hands of the future Minister of Education, and to evade the more troublesome difficulties of the education question by leaving them to be settled between that Minister and the School Boards.

I express no opinion whether it is, or is not, desirable that such powers of controlling all the School Boards in the country should be possessed by a person who may be, like Mr. Forster, eminently likely to use these powers justly and wisely, but who also may be quite the reverse. I merely wish to draw attention to the fact that such powers are given to the Minister, whether he be fit or unfit. The extent of these powers becomes apparent when the other sections of the Act referred to are considered. The fourth clause of the seventh section says :—

"The school shall be conducted in accordance with the conditions required to be fulfilled by an elementary school in order to obtain an annual Parliamentary grant."

What these conditions are appears from the following clauses of the ninety-seventh section :—

"The conditions required to be fulfilled by an elementary school in order to obtain an annual Parliamentary grant shall be those contained in the minutes of the Education Department in force for the time being. . . . Provided that no such minute of the Education Department, not in force at the time of the passing of this Act, shall be deemed to be in force until it has lain for not less than one month on the table of both Houses of Parliament "

Let us consider how this will work in practice. A school established by a School Board may receive support from three sources—from the rates, the school fees, and the Parliamentary grant. The latter may be as great as the two former taken together; and as it may be assumed, without much risk of error, that a constant pressure will be exerted by the ratepayers on the members who represent them to get as much out of the Government, and as little out of the rates, as possible, the School Boards will have a very strong motive for shaping the education they give, as nearly as may be, on the model which the Education Minister offers for their imitation, and for the copying of which he is prepared to pay.

The Revised Code did not compel any school-master to leave off teaching anything; but, by the very simple process of refusing to pay for many kinds of teaching, it has practically put an end to them. Mr. Forster is said to be engaged in revising the Revised Code; a successor of his may re-revise it—and there will be no sort of check upon these revisions and counter revisions, except the possibility of a Parliamentary debate, when the revised, or added, minutes are laid upon the table. What chance is there that any such debate will take place on a matter of detail relating to elementary education—a subject with which members of the Legislature, having been, for the most part, sent to our public schools thirty years ago, have not the least practical acquaintance, and for which they care nothing, unless it derives a political value from its connection with sectarian politics ?

I cannot but think, then, that the School Boards will have the appearance, but not the reality, of freedom of action, in regard to the subject-matter of what is commonly called "secular" education.

As respects what is commonly called "religious" education, the power of the Minister of Education is even more despotic. An interest, almost amounting to pathos, attaches itself, in my mind, to the frantic exertions which are at present going on in almost every school division, to elect certain candidates whose names have never before been heard of in connection with education, and who are either sectarian partisans, or nothing. In my own particular division, a body organised *ad hoc* is moving heaven and earth to get the seven seats filled by seven gentlemen, four of whom are good Churchmen, and three no less good Dissenters. But why should this seven times heated fiery furnace of theological zeal be so desirous to shed its genial warmth over the London School Board ? Can it be that these zealous sectaries mean to evade the solemn pledge given in the Act ?

" No religious catechism or religious formulary which is dis-tinctive of any particular denomination shall be taught in the school."

I confess I should have thought it my duty to reject any such suggestion, as dishonouring to a number of worthy persons, if it had not been for a leading article and some correspondence which appeared in the *Guardian* of November 9th, 1870.

The *Guardian* is, as everybody knows, one of the best of the "religious" newspapers; and, personally, I have every reason to speak highly of the fairness, and indeed kindness, with which the editor is good enough to deal with a writer who must, in many ways, be so objectionable to him as myself. I quote the following passages from a leading article on a letter of mine, therefore, with all respect, and with a genuine conviction that the course of conduct advocated by the writer must appear to him in a very different light from that under which I see it :—

"The first of these points is the interpretation which Professor Huxley puts on the 'Cowper-Temple clause.' It is, in fact, that which we foretold some time ago as likely to be forced upon it by those who think with him. The clause itself was one of those compromises which it is very difficult to define or to maintain logically. On the one side was the simple freedom to School Boards to establish what schools they pleased, which Mr. Forster originally gave, but against which the Nonconformists lifted up their voices, because they conceived it likely to give too much power to the Church. On the other side there was the proposition to make the schools secular—intelligible enough, but in the consideration of public opinion simply impossible—and there was the vague impracticable idea, which Mr. Gladstone thoroughly tore to pieces, of enacting that the teaching of all school-masters in the new schools should be strictly 'undenominational.' The Cowper-Temple clause was, we repeat, proposed simply to tide over the difficulty. It was to satisfy the Nonconformists and the 'unsectarian,' as distinct from the secular party of the League, by forbidding all distinctive 'catechisms and formularies,' which might have the effect of openly assigning the schools to this or that religious body. It refused, at the same time, to attempt the impossible task of defining what was undenominational; and its author even contended, if we understood him correctly, that it would in no way, even indirectly, interfere with the substantial teaching of any master in any school. This assertion we always believed to be untenable; we could not see how, in the face of this clause, a distinctly denominational tone could be honestly given to schools nominally general. But beyond this mere suggestion of an attempt at a general tone of comprehensiveness in religious teaching it was not intended to go, and only because such was its limitation was it accepted by the Government and by the House.

"But now we are told that it is to be construed as doing precisely that which it refused to do. A 'formulary,' it seems, is a collection of formulas, and formulas are simply propositions of whatever kind touching religious faith. All such propositions, if they cannot be accepted by all Christian denominations, are to be proscribed; and it is added significantly that the Jews also are a denomination, and so that any teaching distinctively Christian is perhaps to be excluded, lest it should interfere with their freedom and rights. Are we then to fall back on the simple reading of the letter of the Bible? No! this, it is granted, would be an 'unworthy pretence.' The teacher is to give 'grammatical, geographical, or historical explanations;' but he is to keep clear of 'theology proper,' because, as Professor Huxley takes great pains to prove, there is no theological teaching which is not opposed by some sect or other, from Roman Catholicism on the one hand to Unitarianism on the other. It was not, perhaps, hard to see that this difficulty would be started; and to those who, like Professor Huxley look at it theoretically, without much practical experience of schools, it may appear serious or unanswerable. But there is very little in it practically; when it is faced determinately and handled firmly, it will soon shrink into its true dimensions. The class who are least frightened at it are the school teachers, simply because they know most about it. It is quite clear that the school managers must be cautioned against allowing their schools to be made places of proselytism : but when this is done, the case is simple enough. Leave the masters under this general understanding to teach freely ; if there is ground of complaint, let it be made, but leave the *onus probandi* on the objectors. For extreme peculiarities of belief or unbelief there is the Conscience Clause ; as to the mass of parents, they will be more anxious to have religion taught than afraid of its assuming this or that particular shade. They will trust the school managers and teachers till they have reason to distrust them, and experience has shown that they may trust them safely enough. Any attempt to throw the burden of making the teaching undenominational upon the managers must be sternly resisted : it is simply evading the intentions of the Act in an elaborate attempt to carry them out. We thank Professor Huxley for the warning. To be forewarned is to be forearmed."

A good deal of light seems to me to be thrown on the practical significance of the opinions expressed in the foregoing extract by the following interesting letter, which appeared in the same paper :—

"Sir,—I venture to send to you the substance of a correspondence with the Education Department upon the question of the lawfulness of religious teaching in rate schools under section 14 (2) of the Act. I asked whether the words 'which is distinctive,' &c., taken grammatically as limiting the prohibition of any religious formulary, might be construed as allowing (subject, however, to the other provisions of the Act) any religious formulary common to any two denominations anywhere in England to be taught in such schools ; and if practi-

cally the limit could not be so extended, but would have to be
fixed according to the special circumstances of each district, then
what degree of general acceptance in a district would exempt
such a formulary from the prohibition ?   The answer to this was
as follows :—'It was understood, when clause 14 of the Education Act was discussed in the House of Commons, that, according to a well-known rule of interpreting Acts of Parliament,
"denomination" must be held to include "denominations."
When any dispute is referred to the Education Department
under the last paragraph of section 16, it will be dealt with
according to the circumstances of the case.'

"Upon my asking further if I might hence infer that the lawfulness of teaching any religious formulary in a rate school
would thus depend *exclusively* on local circumstances, and would
accordingly be so decided by the Education Department in case
of dispute, I was informed in explanation that 'their lordships''
letter was intended to convey to me that no general rule, beyond
that stated in the first paragraph of their letter, could at present
be laid down by them ; and that their decision in each particular case must depend on the special circumstances accompanying it.

"I think it would appear from this that it may yet be in
many cases both lawful and expedient to teach religious formularies in rate schools.   H. I.

"Steyning, *November* 5, 1870."

Of course I do not mean to suggest that the
editor of the *Guardian* is bound by the opinions
of his correspondent; but I cannot help thinking
that I do not misrepresent him, when I say that he
also thinks " that it may yet be, in many cases, both
lawful and expedient to teach religious formularies
in rate schools under these circumstances."

It is not uncharitable, therefore, to assume that,
the express words of the Act of Parliament not-

withstanding, all the sectaries who are toiling so hard for seats in the London School Board have the lively hope of the gentleman from Steyning, that it may be "both lawful and expedient to teach religious formularies in rate schools;" and that they mean to do their utmost to bring this happy consummation about.[1]

Now the pathetic emotion to which I have referred, as accompanying my contemplations of the violent struggles of so many excellent persons, is caused by the circumstance that, so far as I can judge, their labour is in vain.

Supposing that the London School Board contains, as it probably will do, a majority of sectaries; and that they carry over the heads of a minority, a resolution that certain theological formulas, about which they all happen to agree,—say, for example, the doctrine of the Trinity,—shall be taught in the schools. Do they fondly imagine that the minority will not at once dispute their interpretation of the

---

[1] A passage in an article on the "Working of the Education Act," in the *Saturday Review* for Nov. 19, 1870, completely justifies this anticipation of the line of action which the sectaries mean to take. After commending the Liverpool compromise, the writer goes on to say :—

"If this plan is fairly adopted in Liverpool, the fourteenth clause of the Act will in effect be restored to its original form, and the majority of the ratepayers in each district be permitted to decide to what denomination the school shall belong."

In a previous paragraph the writer speaks of a possible "mistrust" of one another by the members of the Board, and seems to anticipate "accusations of dishonesty." If any of the members of the Board adopt his views, I think it highly probable that he may turn out to be a true prophet.

Act, and appeal to the Education Department to settle that dispute? And if so, do they suppose that any Minister of Education, who wants to keep his place, will tighten boundaries which the Legislature has left loose; and will give a "final decision" which shall be offensive to every Unitarian and to every Jew in the House of Commons, besides creating a precedent which will afterwards be used to the injury of every Nonconformist? The editor of the *Guardian* tells his friends sternly to resist every attempt to throw the burden of making the teaching undenominational on the managers, and thanks me for the warning I have given him. I return the thanks, with interest, for *his* warning, as to the course the party he represents intends to pursue, and for enabling me thus to draw public attention to a perfectly constitutional and effectual mode of checkmating them.

And, in truth, it is wonderful to note the surprising entanglement into which our able editor gets himself in the struggle between his native honesty and judgment and the necessities of his party. "We could not see," says he, "in the face of this clause how a distinct denominational tone could be honestly given to schools nominally general." There speaks the honest and clear-headed man. "Any attempt to throw the burden of making the teaching undenominational must be sternly resisted." There speaks the advocate holding a brief for his party. "Verily," as Trinculo says, "the monster hath two mouths:" the one, the forward mouth, tells us very justly that the teaching cannot "honestly" be "distinctly denominational;" but the other, the backward mouth, asserts that it must by no manner of means be "undenomina-tional." Putting the two utterances together, I can only interpret them to mean that the teaching is to be "indistinctly denominational." If the editor of the *Guardian* had not shown signs of anger at my use of the term "theological fog," I should have been tempted to suppose it must have been what he had in his mind, under the name of "indistinct denominationalism." But this reading being plainly inadmissible, I can only imagine that he inculcates the teaching of formulas common to a number of denominations.

But the Education Department has already told the gentleman from Steyning that any such pro-ceeding will be illegal. "According to a well-known rule of interpreting Acts of Parliament, 'denom-ination' would be held to include 'denominations.'" In other words, we must read the Act thus :—

"No religious catechism or religious formulary which is distinctive of any particular *denominations* shall be taught."

Thus we are really very much indebted to the editor of the *Guardian* and his correspondent. The one has shown us that the sectaries mean to try to get as much denominational teaching as they can agree upon among themselves, forced into the

c c 2

elementary schools; while the other has obtained
a formal declaration from the Educational Department that any such attempt will contravene the
Act of Parliament, and that, therefore, the unsectarian, law-abiding members of the School Boards
may safely reckon upon bringing down upon their
opponents the heavy hand of the Minister of
Education.[1]

So much for the powers of the School Boards.
Limited as they seem to be, it by no means follows
that such Boards, if they are composed of intelligent and practical men, really more in earnest
about education than about sectarian squabbles,
may not exert a very great amount of influence.
And, from many circumstances, this is especially
likely to be the case with the London School Board,
which, if it conducts itself wisely, may become a
true educational parliament, as subordinate in authority to the Minister of Education, theoretically,
as the Legislature is to the Crown, and yet, like
the Legislature, possessed of great practical
authority. And I suppose that no Minister of
Education would be other than glad to have the

---

[1] Since this paragraph was written, Mr. Forster, in speaking
at the Birkbeck Institution, has removed all doubt as to what his
"final decision" will be in the case of such disputes being
referred to him :—"I have the fullest confidence that in the
reading and explaining of the Bible, what the children will be
taught will be the great truths of Christian life and conduct,
which all of us desire they should know, and that no effort will
be made to cram into their poor little minds, theological dogmas
which their tender age prevents them from understanding."

aid of the deliberations of such a body, or fail to pay careful attention to its recommendations.

What, then, ought to be the nature and scope of the education which a School Board should endeavour to give to every child under its influence, and for which it should try to obtain the aid of the Parliamentary grants ? In my judgment it should include at least the following kinds of instruction and of discipline :—

1. Physical training and drill, as part of the regular business of the school.

It is impossible to insist too much on the importance of this part of education for the children of the poor of great towns. All the conditions of their lives are unfavourable to their physical well-being. They are badly lodged, badly housed, badly fed, and live from one year's end to another in bad air, without chance of a change. They have no play-grounds; they amuse themselves with marbles and chuck-farthing, instead of cricket or hare-and-hounds ; and if it were not for the wonderful instinct which leads all poor children of tender years to run under the feet of cab-horses whenever they can, I know not how they would learn to use their limbs with agility.

Now there is no real difficulty about teaching drill and the simpler kinds of gymnastics. It is done admirably well, for example, in the North Surrey Union schools ; and a year or two ago when I had an opportunity of inspecting these schools, I was greatly struck with the effect of such training upon the poor little waifs and strays of humanity, mostly picked out of the gutter, who are being made into cleanly, healthy, and useful members of society in that excellent institution.

Whatever doubts people may entertain about the efficacy of natural selection, there can be none about artificial selection; and the breeder who should attempt to make, or keep up, a fine stock of pigs, or sheep, under the conditions to which the children of the poor are exposed, would be the laughing-stock even of the bucolic mind. Parliament has already done something in this direction by declining to be an accomplice in the asphyxiation of school children. It refuses to make any grant to a school in which the cubical contents of the school-room are inadequate to allow of proper respiration. I should like to see it make another step in the same direction, and either refuse to give a grant to a school in which physical training is not a part of the programme, or, at any rate, offer to pay upon such training. If something of the kind is not done, the English physique, which has been, and is still, on the whole, a grand one, will become as extinct as the dodo in the great towns.

And then the moral and intellectual effect of drill, as an introduction to, and aid of, all other sorts of training, must not be overlooked. If you want to break in a colt, surely the first thing to do is to catch him and get him quietly to face his trainer; to know his voice and bear his hand; to learn that colts have something else to do with their heels than to kick them up whenever they feel so inclined; and to discover that the dreadful human figure has no desire to devour, or even to beat him, but that, in case of attention and obedience, he may hope for patting and even a sieve of oats.

But, your "street Arabs," and other neglected poor children, are rather worse and wilder than colts; for the reason that the horse-colt has only his animal instincts in him, and his mother, the mare, has been always tender over him, and never came home drunk and kicked him in her life; while the man-colt is inspired by that very real devil, perverted manhood, and *his* mother may have done all that and more. So, on the whole, it may probably be even more expedient to begin your attempt to get at the higher nature of the child, than at that of the colt, from the physical side.

2. Next in order to physical training I put the instruction of children, and especially of girls, in the elements of household work and of domestic economy; in the first place for their own sakes, and in the second for that of their future employers.

Every one who knows anything of the life of the English poor is aware of the misery and waste caused by their want of knowledge of domestic economy, and by their lack of habits of frugality

and method. I suppose it is no exaggeration to
say that a poor Frenchwoman would make the
money which the wife of a poor Englishman
spends in food go twice as far, and at the same
time turn out twice as palatable a dinner. Why
Englishmen, who are so notoriously fond of good
living, should be so helplessly incompetent in the
art of cookery, is one of the great mysteries of
nature ; but from the varied abominations of the
railway refreshment-rooms to the monotonous
dinners of the poor, English feeding is either
wasteful or nasty, or both.

And as to domestic service, the groans of the
housewives of England ascend to heaven ! In five
cases out of six the girl who takes a " place " has
to be trained by her mistress in the first rudiments
of decency and order ; and it is a mercy if she does
not turn up her nose at anything like the mention
of an honest and proper economy. Thousands of
young girls are said to starve, or worse, yearly in
London ; and at the same time thousands of
mistresses of households are ready to pay high
wages for a decent housemaid, or cook, or a fair
workwoman ; and can by no means get what they
want.

Surely, if the elementary schools are worth anything, they may put an end to a state of things
which is demoralising the poor, while it is wasting
the lives of those better off in small worries and
annoyances.

3. But the boys and girls for whose education the School Boards have to provide, have not merely to discharge domestic duties, but each of them is a member of a social and political organisation of great complexity, and has, in future life, to fit himself into that organisation, or be crushed by it. To this end it is surely needful, not only that they should be made acquainted with the elementary laws of conduct, but that their affections should be trained, so as to love with all their hearts that conduct which tends to the attainment of the highest good for themselves and their fellow men, and to hate with all their hearts that opposite course of action which is fraught with evil.

So far as the laws of conduct are determined by the intellect, I apprehend that they belong to science, and to that part of science which is called morality. But the engagement of the affections in favour of that particular kind of conduct which we call good, seems to me to be something quite beyond mere science. And I cannot but think that it, together with the awe and reverence, which have no kinship with base fear, but arise whenever one tries to pierce below the surface of things, whether they be material or spiritual, constitutes all that has any unchangeable reality in religion.

And just as I think it would be a mistake to confound the science, morality, with the affection, religion ; so do I conceive it to be a most lament-able and mischievous error, that the science, theology, is so confounded in the minds of many—indeed, I might say, of the majority of men.

I do not express any opinion as to whether theology is a true science, or whether it does not come under the apostolic definition of " science falsely so called ;" though I may be permitted to express the belief that if the Apostle to whom that much misapplied phrase is due could make the acquaintance of much of modern theology, he would not hesitate a moment in declaring that it is exactly what he meant the words to denote.

But it is at any rate conceivable, that the nature of the Deity, and his relations to the universe, and more especially to mankind, are capable of being ascertained, either inductively or deductively, or by both processes. And, if they have been ascertained, then a body of science has been formed which is very properly called theology.

Further, there can be no doubt that affection for the Being thus defined and described by theologic science would be properly termed re-ligion ; but it would not be the whole of religion. The affection for the ethical ideal defined by moral science would claim equal if not superior rights. For suppose theology established the existence of an evil deity—and some theologies, even Christian ones, have come very near this,—

is the religious affection to be transferred from the ethical ideal to any such omnipotent demon ? I trow not. Better a thousand times that the human race should perish under his thunderbolts than it should say, " Evil, be thou my good."

There is nothing new, that I know of, in this statement of the relations of religion with the science of morality on the one hand and that of theology on the other. But I believe it to be altogether true, and very needful, at this time, to be clearly and emphatically recognised as such, by those who have to deal with the education question.

We are divided into two parties—the advocates of so-called " religious " teaching on the one hand, and those of so-called "secular" teaching on the other. And both parties seem to me to be not only hopelessly wrong, but in such a position that if either succeeded completely, it would discover, before many years were over, that it had made a great mistake and done serious evil to the cause of education.

For, leaving aside the more far-seeing minority on each side, what the "religious" party is crying for is mere theology, under the name of religion ; while the "secularists" have unwisely and wrongfully admitted the assumption of their opponents, and demand the abolition of all " religious" teaching, when they only want to be free of theology— Burning your ship to get rid of the cockroaches !

But my belief is, that no human being, and no society composed of human beings, ever did, or ever will, come to much, unless their conduct was governed and guided by the love of some ethical ideal. Undoubtedly, your gutter child may be converted by mere intellectual drill into "the subtlest of all the beasts of the field;" but we know what has become of the original of that description, and there is no need to increase the number of those who imitate him successfully without being aided by the rates. And if I were compelled to choose for one of my own children, between a school in which real religious instruction is given, and one without it, I should prefer the former, even though the child might have to take a good deal of theology with it. Nine-tenths of a dose of bark is mere half-rotten wood; but one swallows it for the sake of the particles of quinine, the beneficial effect of which may be weakened, but is not destroyed, by the wooden dilution, unless in a few cases of exceptionally tender stomachs.

Hence, when the great mass of the English people declare that they want to have the children in the elementary schools taught the Bible, and when it is plain from the terms of the Act, the debates in and out of Parliament, and especially the emphatic declarations of the Vice-President of the Council, that it was intended that such Bible-reading should be permitted, unless good cause for prohibiting it could be shown, I do not see what reason there is for opposing that wish. Certainly, I, individually, could with no shadow of consistency oppose the teaching of the children of other people to do that which my own children are taught to do. And, even if the reading the Bible were not, as I think it is, consonant with political reason and justice, and with a desire to act in the spirit of the education measure, I am disposed to think it might still be well to read that book in the elementary schools.

I have always been strongly in favour of secular education, in the sense of education without theology ; but I must confess I have been no less seriously perplexed to know by what practical measures the religious feeling, which is the essential basis of conduct, was to be kept up, in the present utterly chaotic state of opinion on these matters, without the use of the Bible. The Pagan moralists lack life and colour, and even the noble Stoic, Marcus Antonius, is too high and refined for an ordinary child. Take the Bible as a whole ; make the severest deductions which fair criticism can dictate for shortcomings and positive errors ; eliminate, as a sensible lay-teacher would do, if left to himself, all that it is not desirable for children to occupy themselves with ; and there still remains in this old literature a vast residuum of moral beauty and grandeur. And then consider the great historical fact that, for three centuries,

this book has been woven into the life of all that
is best and noblest in English history; that it has
become the national epic of Britain, and is as
familiar to noble and simple, from John-o'-Groat's
House to Land's End, as Dante and Tasso once
were to the Italians; that it is written in the
noblest and purest English, and abounds in ex-
quisite beauties of mere literary form; and,
finally, that it forbids the veriest hind who never
left his village to be ignorant of the existence of
other countries and other civilisations, and of a
great past, stretching back to the furthest limits
of the oldest nations in the world.   By the study
of what other book could children be so much
humanised and made to feel that each figure in
that vast historical procession fills, like themselves,
but a momentary space in the interval between
two eternities; and earns the blessings or the curses
of all time, according to its effort to do good and
hate evil, even as they also are earning their payment for their work ?

On the whole, then, I am in favour of reading
the Bible, with such grammatical, geographical,
and historical explanations by a lay-teacher as may
be needful, with rigid exclusion of any further
theological teaching than that contained in the
Bible itself.   And in stating what this is, the
teacher would do well not to go beyond the
precise words of the Bible; for if he does, he will,
in the first place, undertake a task beyond his

strength, seeing that all the Jewish and Christian sects have been at work upon that subject for more than two thousand years, and have not yet arrived, and are not in the least likely to arrive, at an agreement ; and, in the second place, he will certainly begin to teach something distinctively denominational, and thereby come into violent collision with the Act of Parliament.

4. The intellectual training to be given in the elementary schools must of course, in the first place, consist in learning to use the means of acquiring knowledge, or reading, writing, and arithmetic; and it will be a great matter to teach reading so completely that the act shall have become easy and pleasant. If reading remains " hard," that accomplishment will not be much resorted to for instruction, and still less for amusement—which last is one of its most valuable uses to hard-worked people. But along with a due proficiency in the use of the means of learning, a certain amount of knowledge, of intellectual discipline, and of artistic training should be conveyed in the elementary schools ; and in this direction—for reasons which I am afraid to repeat, having urged them so often—I can conceive no subject-matter of education so appropriate and so important as the rudiments of physical science, with drawing, modelling, and singing. Not only would such teaching afford the best possible preparation for the technical schools

about which so much is now said, but the organisation for carrying it into effect already exists. The Science and Art Department, the operations of which have already attained considerable magnitude, not only offers to examine and pay the results of such examination in elementary science and art, but it provides what is still more important, viz. a means of giving children of high natural ability, who are just as abundant among the poor as among the rich, a helping hand. A good old proverb tells us that " One should not take a razor to cut a block : " the razor is soon spoiled, and the block is not so well cut as it would be with a hatchet. But it is worse economy to prevent a possible Watt from being anything but a stoker, or to give a possible Faraday no chance of doing anything but to bind books. Indeed, the loss in such cases of mistaken vocation has no measure; it is absolutely infinite and irreparable. And among the arguments in favour of the interference of the State in education, none seems to be stronger than this—that it is the interest of every one that ability should be neither wasted, nor misapplied, by any one : and, therefore, that every one's representative, the State, is necessarily fulfilling the wishes of its constituents when it is helping the capacities to reach their proper places.

It may be said that the scheme of education here sketched is too large to be effected in the time during which the children will remain at school; and, secondly, that even if this objection did not exist, it would cost too much.

I attach no importance whatever to the first objection until the experiment has been fairly tried. Considering how much catechism, lists of the kings of Israel, geography of Palestine, and the like, children are made to swallow now, I cannot believe there will be any difficulty in inducing them to go through the physical training, which is more than half play; or the instruction in household work, or in those duties to one another and to themselves, which have a daily and hourly practical interest. That children take kindly to elementary science and art no one can doubt who has tried the experiment properly. And if Bible-reading is not accompanied by constraint and solemnity, as if it were a sacramental operation, I do not believe there is anything in which children take more pleasure. At least I know that some of the pleasantest recollections of my childhood are connected with the voluntary study of an ancient Bible which belonged to my grand-mother. There were splendid pictures in it, to be sure; but I recollect little or nothing about them save a portrait of the high priest in his vestments. What come vividly back on my mind are remembrances of my delight in the histories of Joseph and of David; and of my keen appreciation of the chivalrous kindness of Abraham in his dealing

with Lot. Like a sudden flash there returns back
upon me, my utter scorn of the pettifogging mean-
ness of Jacob, and my sympathetic grief over the
heartbreaking lamentation of the cheated Esau,
" Hast thou not a blessing for me also, O my
father ? " And I see, as in a cloud, pictures
of the grand phantasmagoria of the Book of Revelation.

I enumerate, as they issue, the childish impressions which come crowding out of the pigeon-holes
in my brain, in which they have lain almost
undisturbed for forty years. I prize them as an
evidence that a child of five or six years old, left
to his own devices, may be deeply interested in
the Bible, and draw sound moral sustenance from it.
And I rejoice that I was left to deal with the
Bible alone ; for if I had had some theological
" explainer " at my side, he might have tried, as
such do, to lessen my indignation against Jacob,
and thereby have warped my moral sense for ever ;
while the great apocalyptic spectacle of the ultimate triumph of right and justice might have been
turned to the base purposes of a pious lampooner
of the Papacy.

And as to the second objection—costliness—
the reply is, first, that the rate and the Parliamentary grant together ought to be enough, considering that science and art teaching is already
provided for ; and, secondly, that if they are not,
it may be well for the educational parliament to

consider what has become of those endowments which were originally intended to be devoted, more or less largely, to the education of the poor.

When the monasteries were spoiled, some of their endowments were applied to the foundation of cathedrals; and in all such cases it was ordered that a certain portion of the endowment should be applied to the purposes of education. How much is so applied? Is that which may be so applied given to help the poor, who cannot pay for education, or does it virtually subsidise the comparatively rich, who can? How are Christ's Hospital and Alleyn's foundation securing their right purposes, or how far are they perverted into contrivances for affording relief to the classes who can afford to pay for education? How— But this paper is already too long, and, if I begin, I may find it hard to stop asking questions of this kind, which after all are worthy only of the lowest of Radicals.

XVI

TECHNICAL EDUCATION.

[1877]

ANY candid observer of the phenomena of modern society will readily admit that bores must be classed among the enemies of the human race; and a little consideration will probably lead him to the further admission, that no species of that extensive genus of noxious creatures is more objectionable than the educational bore. Convinced as I am of the truth of this great social generalisation, it is not without a certain trepidation that I venture to address you on an educational topic. For, in the course of the last ten years, to go back no farther, I am afraid to say how often I have ventured to speak of education, from that given in the primary schools to that which is to be had in the universities and medical colleges; indeed, the only part of this wide region into which, as yet, I have not adventured is that into which I propose to intrude to-day.

Thus, I cannot but be aware that I am danger-
ously near becoming the thing which all men fear
and fly. But I have deliberately elected to run
the risk. For when you did me the honour to ask
me to address you, an unexpected circumstance had
led me to occupy myself seriously with the question
of technical education; and I had acquired the conviction that there are few subjects respecting which
it is more important for all classes of the community to have clear and just ideas than this; while,
certainly, there is none which is more deserving
of attention by the Working Men's Club and
Institute Union.

It is not for me to express an opinion whether
the considerations, which I am about to submit to
you, will be proved by experience to be just or not,
but I will do my best to make them clear. Among
the many good things to be found in Lord Bacon's
works, none is more full of wisdom than the saying
that "truth more easily comes out of error than
out of confusion." Clear and consecutive wrong-
thinking is the next best thing to right-thinking;
so that, if I succeed in clearing your ideas on this
topic, I shall have wasted neither your time nor
my own.

" Technical education," in the sense in which the
term is ordinarily used, and in which I am now
employing it, means that sort of education which
is specially adapted to the needs of men whose
business in life it is to pursue some kind of handi-

craft; it is, in fact, a fine Greco-Latin equivalent
for what in good vernacular English would be
called "the teaching of handicrafts."   And probably, at this stage of our progress, it may occur to
many of you to think of the story of the cobbler and
his last, and to say to yourselves, though you will
be too polite to put the question openly to me,
What does the speaker know practically about this
matter?   What is his handicraft?   I think the
question is a very proper one, and unless I were
prepared to answer it, I hope satisfactorily, I
should have chosen some other theme.

   The fact is, I am, and have been, any time these
thirty years, a man who works with his hands—a
handicraftsman.   I do not say this in the broadly
metaphorical sense in which fine gentlemen, with
all the delicacy of Agag about them, trip to the
hustings about election time, and protest that they
too are working men.   I really mean my words to
be taken in their direct, literal, and straightforward
sense.   In fact, if the most nimble-fingered watchmaker among you will come to my workshop, he
may set me to put a watch together, and I will set
him to dissect, say, a blackbeetle's nerves.   I do
not wish to vaunt, but I am inclined to think
that I shall manage my job to his satisfaction
sooner than he will do his piece of work to mine.

   In truth, anatomy, which is my handicraft, is
one of the most difficult kinds of mechanical labour,
involving, as it does, not only lightness and dex-

terity of hand, but sharp eyes and endless patience. And you must not suppose that my particular branch of science is especially distinguished for the demand it makes upon skill in manipulation. A similar requirement is made upon all students of physical science. The astronomer, the electrician, the chemist, the mineralogist, the botanist, are constantly called upon to perform manual operations of exceeding delicacy. The progress of all branches of physical science depends upon observation, or on that artificial observation which is termed experiment, of one kind or another; and, the farther we advance, the more practical difficulties surround the investigation of the conditions of the problems offered to us; so that mobile and yet steady hands, guided by clear vision, are more and more in request in the workshops of science.

Indeed, it has struck me that one of the grounds of that sympathy between the handicraftsmen of this country and the men of science, by which it has so often been my good fortune to profit, may, perhaps, lie here. You feel and we feel that, among the so-called learned folks, we alone are brought into contact with tangible facts in the way that you are. You know well enough that it is one thing to write a history of chairs in general, or to address a poem to a throne, or to speculate about the occult powers of the chair of St. Peter; and quite another thing to make with your own hands a veritable chair, that will stand fair and square,

And now, having, as I hope, justified my assumption of a place among handicraftsmen, and put myself right with you as to my qualification, from practical knowledge, to speak about technical education, I will proceed to lay before you the results of my experience as a teacher of a handicraft, and tell you what sort of education I should think best adapted for a boy whom one wanted to make a professional anatomist.

I should say, in the first place, let him have a good English elementary education. I do not mean that he shall be able to pass in such and such a standard—that may or may not be an equivalent expression—but that his teaching shall have been such as to have given him command of the common implements of learning and to have created a desire for the things of the understanding.

Further, I should like him to know the elements of physical science, and especially of physics and chemistry, and I should take care that this elementary knowledge was real. I should like my aspirant to be able to read a scientific treatise in Latin, French, or German, because an enormous amount of anatomical knowledge is locked up in those languages. And especially, I should require some ability to draw—I do not mean artistically, for that is a gift which may be cultivated but cannot be learned, but with fair accuracy. I will not say that everybody can learn even this; for the negative development of the faculty of drawing in some people is almost miraculous. Still everybody, or almost everybody, can learn to write; and, as writing is a kind of drawing, I suppose that the majority of the people who say they cannot draw, and give copious evidence of the accuracy of their assertion, could draw, after a fashion, if they tried. And that "after a fashion" would be better than nothing for my purposes.

Above all things, let my imaginary pupil have preserved the freshness and vigour of youth in his mind as well as his body. The educational abomination of desolation of the present day is the stimulation of young people to work at high pressure by incessant competitive examinations. Some wise man (who probably was not an early riser) has said of early risers in general, that they are conceited all the forenoon and stupid all the afternoon. Now whether this is true of early risers in the common acceptation of the word or not, I will not pretend to say ; but it is too often true of the unhappy children who are forced to rise too early in their classes. They are conceited all the forenoon of life, and stupid all its afternoon. The vigour and freshness, which should have been stored up for the purposes of the hard struggle for existence in practical life, have been washed out of them by precocious mental debauchery—by book gluttony and lesson bibbing. Their faculties are worn out by the strain put upon their

callow brains, and they are demoralised by worthless childish triumphs before the real work of life
begins. I have no compassion for sloth, but youth
has more need for intellectual rest than age; and
the cheerfulness, the tenacity of purpose, the power
of work which make many a successful man what
he is, must often be placed to the credit, not of
his hours of industry, but to that of his hours of
idleness, in boyhood. Even the hardest worker
of us all, if he has to deal with anything above
mere details, will do well, now and again, to let
his brain lie fallow for a space. The next crop of
thought will certainly be all the fuller in the ear
and the weeds fewer.

This is the sort of education which I should
like any one who was going to devote himself to
my handicraft to undergo. As to knowing anything about anatomy itself, on the whole I would
rather he left that alone until he took it up
seriously in my laboratory. It is hard work
enough to teach, and I should not like to have
superadded to that the possible need of un-
teaching.

Well, but, you will say, this is Hamlet with the
Prince of Denmark left out; your "technical
education" is simply a good education, with
more attention to physical science, to drawing, and to modern languages than is common, and there is nothing specially technical
about it.

Exactly so; that remark takes us straight to the heart of what I have to say; which is, that, in my judgment, the preparatory education of the handicraftsman ought to have nothing of what is ordinarily understood by "technical" about it.

The workshop is the only real school for a handicraft. The education which precedes that of the workshop should be entirely devoted to the strengthening of the body, the elevation of the moral faculties, and the cultivation of the intelligence; and, especially, to the imbuing the mind with a broad and clear view of the laws of that natural world with the components of which the handicraftsman will have to deal. And, the earlier the period of life at which the handicraftsman has to enter into actual practice of his craft, the more important is it that he should devote the precious hours of preliminary education to things of the mind, which have no direct and immediate bearing on his branch of industry, though they lie at the foundation of all realities.

Now let me apply the lessons I have learned from my handicraft to yours. If any of you were obliged to take an apprentice, I suppose you would like to get a good healthy lad, ready and willing to learn, handy, and with his fingers not all thumbs, as the saying goes. You would like that he should read, write, and cipher well; and, if you were an intelligent master, and your trade involved the application of scientific principles, as so many trades do, you would like him to know enough of the elementary principles of science to understand what was going on. I suppose that, in nine trades out of ten, it would be useful if he could draw ; and many of you must have lamented your inability to find out for yourselves what foreigners are doing or have done. So that some knowledge of French and German might, in many cases, be very desirable.

So it appears to me that what you want is pretty much what I want ; and the practical question is, How you are to get what you need, under the actual limitations and conditions of life of handicraftsmen in this country ?

I think I shall have the assent both of the employers of labour and of the employed as to one of these limitations ; which is, that no scheme of technical education is likely to be seriously entertained which will delay the entrance of boys into working life, or prevent them from contributing towards their own support, as early as they do at present. Not only do I believe that any such scheme could not be carried out, but I doubt its desirableness, even if it were practicable.

The period between childhood and manhood is full of difficulties and dangers, under the most favourable circumstances ; and, even among the well-to-do, who can afford to surround their children with the most favourable conditions, examples of a career ruined, before it has well begun, are but too frequent. Moreover, those who have to live by labour must be shaped to labour early. The colt that is left at grass too long makes but a sorry draught-horse, though his way of life does not bring him within the reach of artificial temptations. Perhaps the most valuable result of all education is the ability to make yourself do the thing you have to do, when it ought to be done, whether you like it or not; it is the first lesson that ought to be learned; and, however early a man's training begins, it is probably the last lesson that he learns thoroughly.

There is another reason, to which I have already adverted, and which I would reiterate, why any extension of the time devoted to ordinary school-work is undesirable. In the newly-awakened zeal for education, we run some risk of forgetting the truth that while under-instruction is a bad thing, over-instruction may possibly be a worse.

Success in any kind of practical life is not dependent solely, or indeed chiefly, upon knowledge. Even in the learned professions, knowledge alone, is of less consequence than people are apt to suppose. And, if much expenditure of bodily energy is involved in the day's work, mere knowledge is of still less importance when weighed against the probable cost of its acquirement. To do a fair day's work with his hands, a man needs, above all things, health, strength, and the patience and cheerfulness which, if they do not always accompany these blessings, can hardly in the nature of things exist without them; to which we must add honesty of purpose and a pride in doing what is done well.

A good handicraftsman can get on very well without genius, but he will fare badly without a reasonable share of that which is a more useful possession for workaday life, namely, mother-wit; and he will be all the better for a real knowledge, however limited, of the ordinary laws of nature, and especially of those which apply to his own business.

Instruction carried so far as to help the scholar to turn his store of mother-wit to account, to acquire a fair amount of sound elementary knowledge, and to use his hands and eyes; while leaving him fresh, vigorous, and with a sense of the dignity of his own calling, whatever it may be, if fairly and honestly pursued, cannot fail to be of invaluable service to all those who come under its influence.

But, on the other hand, if school instruction is carried so far as to encourage bookishness; if the ambition of the scholar is directed, not to the gaining of knowledge, but to the being able to pass examinations successfully; especially if encouragement is given to the mischievous delusion that brainwork is, in itself, and apart from its quality, a nobler or more respectable thing than handiwork

—such education may be a deadly mischief to the workman, and lead to the rapid ruin of the industries it is intended to serve.

I know that I am expressing the opinion of some of the largest as well as the most enlightened employers of labour, when I say that there is a real danger that, from the extreme of no education, we may run to the other extreme of over-education of handicraftsmen. And I apprehend that what is true for the ordinary hand-worker is true for the foreman. Activity, probity, knowledge of men, ready mother-wit, supplemented by a good knowledge of the general principles involved in his business, are the making of a good foreman. If he possess these qualities, no amount of learning will fit him better for his position; while the course of life and the habit of mind required for the attainment of such learning may, in various direct and indirect ways, act as direct disqualifications for it.

Keeping in mind, then, that the two things to be avoided are, the delay of the entrance of boys into practical life, and the substitution of exhausted bookworms for shrewd, handy men, in our works and factories, let us consider what may be wisely and safely attempted in the way of improving the education of the handicraftsman.

First, I look to the elementary schools now happily established all over the country. I am not going to criticise or find fault with them; on the contrary, their establishment seems to me to be the most important and the most beneficial result of the corporate action of the people in our day. A great deal is said of British interests just now, but, depend upon it, that no Eastern difficulty needs our intervention as a nation so seriously, as the putting down both the Bashi-Bazouks of ignorance and the Cossacks of sectarianism at home. What has already been achieved in these directions is a great thing; you must have lived some time to know how great. An education, better in its processes, better in its substance, than that which was accessible to the great majority of well-to-do Britons a quarter of a century ago, is now obtainable by every child in the land. Let any man of my age go into an ordinary elementary school, and unless he was unusually fortunate in his youth, he will tell you that the educational method, the intelligence, patience, and good temper on the teacher's part, which are now at the disposal of the veriest waifs and wastrels of society, are things of which he had no experience in those costly, middle-class schools, which were so ingeniously contrived as to combine all the evils and short-comings of the great public schools with none of their advantages. Many a man, whose so-called education cost a good deal of valuable money and occupied many a year of invaluable time, leaves the inspection of a well-ordered elementary school devoutly wishing that, in his young days, he had ant branches of education, nothing more than the rudiments of science and art teaching can be introduced into elementary schools, we must seek elsewhere for a supplementary training in these subjects, and, if need be, in foreign languages, which may go on after the workman's life has begun.

The means of acquiring the scientific and artistic part of this training already exists in full working order, in the first place, in the classes of the Science and Art Department, which are, for the most part, held in the evening, so as to be accessible to all who choose to avail themselves of them after working hours. The great advantage of these classes is that they bring the means of instruction to the doors of the factories and work-shops ; that they are no artificial creations, but by their very existence prove the desire of the people for them ; and finally, that they admit of indefinite development in proportion as they are wanted. I have often expressed the opinion, and I repeat it here, that, during the eighteen years they have been in existence these classes have done incalculable good ; and I can say, of my own know-ledge, that the Department spares no pains and trouble in trying to increase their usefulness and ensure the soundness of their work.

No one knows better than my friend Colonel Donnelly, to whose clear views and great adminis-trative abilities so much of the successful working

of the science classes is due, that there is much to
be done before the system can be said to be
thoroughly satisfactory. The instruction given
needs to be made more systematic and especially
more practical; the teachers are of very unequal
excellence, and not a few stand much in need of
instruction themselves, not only in the subject
which they teach, but in the objects for which
they teach. I dare say you have heard of that
proceeding, reprobated by all true sportsmen,
which is called "shooting for the pot." Well,
there is such a thing as "teaching for the pot"—
teaching, that is, not that your scholar may know,
but that he may count for payment among those
who pass the examination; and there are some
teachers, happily not many, who have yet to learn
that the examiners of the Department regard them
as poachers of the worst description.

Without presuming in any way to speak in the
name of the Department, I think I may say, as a
matter which has come under my own observation,
that it is doing its best to meet all these difficulties.
It systematically promotes practical instruction in
the classes; it affords facilities to teachers who
desire to learn their business thoroughly; and it
is always ready to aid in the suppression of pot-
teaching.

All this is, as you may imagine, highly satisfactory to me. I see that spread of scientific
education, about which I have so often permitted

myself to worry the public, become, for all practical purposes, an accomplished fact. Grateful as I am for all that is now being done, in the same direction, in our higher schools and universities, I have ceased to have any anxiety about the wealthier classes. Scientific knowledge is spreading by what the alchemists called a "distillatio per ascensum ;" and nothing now can prevent it from continuing to distil upwards and permeate English society, until, in the remote future, there shall be no member of the legislature who does not know as much of science as an elementary school-boy ; and even the heads of houses in our venerable seats of learning shall acknowledge that natural science is not merely a sort of University back-door through which inferior men may get at their degrees. Perhaps this apocalyptic vision is a little wild ; and I feel I ought to ask pardon for an outbreak of enthusiasm, which, I assure you, is not my commonest failing.

I have said that the Government is already doing a great deal in aid of that kind of technical education for handicraftsmen which, to my mind, is alone worth seeking. Perhaps it is doing as much as it ought to do, even in this direction. Certainly there is another kind of help of the most important character, for which we may look elsewhere than to the Government. The great mass of mankind have neither the liking, nor the aptitude, for either literary, or scientific, or artistic pursuits ; nor,

indeed, for excellence of any sort.   Their ambition
is to go through life with moderate exertion and a
fair share of ease, doing common things in a common way.   And a great blessing and comfort it is
that the majority of men are of this mind ; for the
majority of things to be done are common things,
and are quite well enough done when commonly
done.   The great end of life is not knowledge but
action.   What men need is, as much knowledge as
they can assimilate and organise into a basis for
action;   give them more and it may become
injurious.   One knows people who are as heavy
and stupid from undigested learning as others are
from over-fulness of meat and drink.   But a small
percentage of the population is born with that most
excellent quality, a desire for excellence, or with
special aptitudes of some sort or another ; Mr.
Galton tells us that not more than one in four thousand may be expected to attain distinction, and not
more than one in a million some share of that
intensity of instinctive aptitude, that burning
thirst for excellence, which is called genius.

Now, the most important object of all educational schemes is to catch these exceptional people,
and turn them to account for the good of society.
No man can say where they will crop up ; like
their opposites, the fools and knaves, they appear
sometimes in the palace, and sometimes in the
hovel ; but the great thing to be aimed at, I was
almost going to say the most important end of all

social arrangements, is to keep these glorious sports of Nature from being either corrupted by luxury or starved by poverty, and to put them into the position in which they can do the work for which they are especially fitted.

Thus, if a lad in an elementary school showed signs of special capacity, I would try to provide him with the means of continuing his education after his daily working life had begun ; if in the evening classes he developed special capabilities in the direction of science or of drawing, I would try to secure him an apprenticeship to some trade in which those powers would have applicability. Or, if he chose to become a teacher, he should have the chance of so doing. Finally, to the lad of genius, the one in a million, I would make accessible the highest and most complete training the country could afford. Whatever that might cost, depend upon it the investment would be a good one. I weigh my words when I say that if the nation could purchase a potential Watt, or Davy, or Faraday, at the cost of a hundred thousand pounds down, he would be dirt-cheap at the money. It is a mere commonplace and everyday piece of knowledge, that what these three men did has produced untold millions of wealth, in the narrowest economical sense of the word.

Therefore, as the sum and crown of what is to be done for technical education, I look to the provision of a machinery for winnowing out the capacities and giving them scope. When I was a member of the London School Board, I said, in the course of a speech, that our business was to provide a ladder, reaching from the gutter to the university, along which every child in the three kingdoms should have the chance of climbing as far as he was fit to go. This phrase was so much bandied about at the time, that, to say truth, I am rather tired of it; but I know of no other which so fully expresses my belief, not only about education in general, but about technical education in particular.

The essential foundation of all the organisation needed for the promotion of education among handicraftsmen will, I believe, exist in this country, when every working lad can feel that society has done as much as lies in its power to remove all needless and artificial obstacles from his path; that there is no barrier, except such as exists in the nature of things, between himself and whatever place in the social organisation he is fitted to fill; and, more than this, that, if he has capacity and industry, a hand is held out to help him along any path which is wisely and honestly chosen.

I have endeavoured to point out to you that a great deal of such an organisation already exists; and I am glad to be able to add that there is a good prospect that what is wanting will, before long, be supplemented.

Those powerful and wealthy societies, the livery companies of the City of London, remembering that they are the heirs and representatives of the trade guilds of the Middle Ages, are interesting themselves in the question. So far back as 1872 the Society of Arts organised a system of instruction in the technology of arts and manufactures, for persons actually employed in factories and workshops, who desired to extend and improve their knowledge of the theory and practice of their particular avocations;[1] and a considerable subsidy, in aid of the efforts of the Society, was liberally granted by the Clothworkers' Company. We have here the hopeful commencement of a rational organisation for the promotion of excellence among handicraftsmen. Quite recently, other of the livery companies have determined upon giving their powerful, and, indeed, almost boundless, aid to the improvement of the teaching of handicrafts. They have already gone so far as to appoint a committee to act for them; and I betray no confidence in adding that, some time since, the committee sought the advice and assistance of several persons, myself among the number.

Of course I cannot tell you what may be the result of the deliberations of the committee; but we may all fairly hope that, before long, steps which will have a weighty and a lasting influence on the growth and

[1] See the *Programme* for 1878, issued by the Society of Arts, p. 14.

XVII

ADDRESS ON BEHALF OF THE NATIONAL ASSOCIATION FOR THE PRO-MOTION OF TECHNICAL EDUCATION

[1887.]

MR. MAYOR AND GENTLEMEN,—It must be a matter of sincere satisfaction to those who, like myself, have for many years past been convinced of the vital importance of technical education to this country to see that that subject is now being taken up by some of the most important of our manufacturing towns. The evidence which is afforded of the public interest in the matter by such meetings as those at Liverpool and New-castle, and, last but not least, by that at which I have the honour to be present to-day, may con-vince us all, I think, that the question has passed out of the region of speculation into that of action. I need hardly say to any one here that the task which our Association contemplates is not only one of primary importance—I may say of vital importance—to the welfare of the country ; but that it is one of great extent and of vast difficulty. There is a well-worn adage that those who set out upon a great enterprise would do well to count the cost. I am not sure that this is always true. I think that some of the very greatest enterprises in this world have been carried out successfully simply because the people who undertook them did not count the cost; and I am much of opinion that, in this very case, the most instructive consideration for us is the cost of doing nothing. But there is one thing that is perfectly certain, and it is that, in undertaking all enterprises, one of the most important conditions of success is to have a perfectly clear comprehension of what you want to do—to have that before your minds before you set out, and from that point of view to consider carefully the measures which are best adapted to the end.

Mr. Acland has just given you an excellent account of what is properly and strictly understood by technical education; but I venture to think that the purpose of this Association may be stated in somewhat broader terms, and that the object we have in view is the development of the industrial productivity of the country to the uttermost limits consistent with social welfare. And you will observe that, in thus widening the definition of our object, I have gone no further than the Mayor in his speech, when he not obscurely hinted—and most justly hinted—that in dealing with this question there are other matters than technical education, in the strict sense, to be considered.

It would be extreme presumption on my part if I were to attempt to tell an audience of gentlemen intimately acquainted with all branches of industry and commerce, such as I see before me, in what manner the practical details of the operations that we propose are to be carried out. I am absolutely ignorant both of trade and of commerce, and upon such matters I cannot venture to say a solitary word. But there is one direction in which I think it possible I may be of· service—not much perhaps, but still of some,—because this matter, in the first place, involves the consideration of methods of education with which it has been my business to occupy myself during the greater part of my life; and, in the second place, it involves attention to some of those broad facts and laws of nature with which it has been my business to acquaint myself to the best of my ability. And what I think may be possible is this, that if I succeed in putting before you—as briefly as I can, but in clear and connected shape—what strikes me as the programme that we have eventually to carry out, and what are the indispensable conditions of success, that that proceeding, whether the conclusions at which I arrive be such as you approve or as you disapprove, will nevertheless help to clear the course.

organisation of this system in London, and I am
glad to think that, after all these years, I can look
back upon that period of my life as perhaps the
part of it least wasted.

No one can doubt that this system of primary
education has done wonders for our population;
but, from our point of view, I do not think anybody can doubt that it still has very considerable
defects. It has the defect which is common to all
the educational systems which we have inherited—
it is too bookish, too little practical. The child is
brought too little into contact with actual facts and
things, and as the system stands at present it constitutes next to no education of those particular
faculties which are of the utmost importance to
industrial life—I mean the faculty of observation,
the faculty of working accurately, of dealing with
things instead of with words. I do not propose to
enlarge upon this topic, but I would venture to
suggest that there are one or two remedial measures
which are imperatively needed; indeed, they have
already been alluded to by Mr. Acland. Those
which strike me as of the greatest importance are
two, and the first of them is the teaching of drawing. In my judgment, there is no mode of
exercising the faculty of observation and the
faculty of accurate reproduction of that which is
observed, no discipline which so readily tests error
in these matters, as drawing properly taught.
And by that I do not mean artistic drawing; I

mean figuring natural objects : making plans and
sections, approaching geometrical rather than
artistic drawing. I do not wish to exaggerate,
but I declare to you that, in my judgment, the
child who has been taught to make an accurate
elevation, plan and section of a pint pot has had
an admirable training in accuracy of eye and hand.
I am not talking about artistic education. That is
not the question. Accuracy is the foundation of
everything else, and instruction in artistic drawing
is something which may be put off till a later
stage. Nothing has struck me more in the course
of my life than the loss which persons, who are
pursuing scientific knowledge of any kind, sustain
from the difficulties which arise because they
never have been taught elementary drawing;
and I am glad to say that in Eton, a school of
whose governing body I have the honour of being
a member, we some years ago made drawing imperative on the whole school.

The other matter in which we want some
systematic and good teaching is what I have
hardly a name for, but which may best be explained as a sort of developed object lessons such
as Mr. Acland adverted to. Anybody who knows
his business in science can make anything subservient to that purpose. You know it was said of
Dean Swift that he could write an admirable poem
upon a broomstick, and the man who has a real
knowledge of science can make the commonest ob-

ject in the world subservient to an introduction to the principles and greater truths of natural knowledge. It is in that way that your science must be taught if it is to be of real service. Do not suppose any amount of book work, any repetition by rote of catechisms and other abominations of that kind are of value for our object. That is mere wasting of time. But take the commonest object and lead the child from that foundation to such truths of a higher order as may be within his grasp. With regard to drawing, I do not think there is any practical difficulty ; but in respect to the scientific object lessons you want teachers trained in a manner different from that which now prevails.

If it is found practicable to add further training of the hand and eye by instruction in modelling or in simple carpentry, well and good. But I should stop at this point. The elementary schools are already charged with quite as much as they can do properly ; and I do not believe that any good can come of burdening them with special technical instruction. Out of that, I think, harm would come.

Now let me pass to my second point, which is the development of technical skill. Everybody here is aware that at this present moment there is hardly a branch of trade or of commerce which does not depend, more or less directly, upon some department or other of physical science, which does not involve, for its successful pursuit, reasoning from scientific data. Our machinery, our chemical processes or dyeworks, and a thousand operations which it is not necessary to mention, are all directly and immediately connected with science. You have to look among your workmen and foremen for persons who shall intelligently grasp the modifications, based upon science, which are constantly being introduced into these industrial processes. I do not mean that you want professional chemists, or physicists, or mathematicians, or the like, but you want people sufficiently familiar with the broad principles which underlie industrial operations to be able to adapt themselves to new conditions. Such qualifications can only be secured by a sort of scientific instruction which occupies a midway place between those primary notions given in the elementary schools and those more advanced studies which would be carried out in the technical schools.

You are aware that, at present, a very large machinery is in operation for the purpose of giving this instruction. I don't refer merely to such work as is being done at Owens College here, for example, or at other local colleges. I allude to the larger operations of the Science and Art Department, with which I have been connected for a great many years. I constantly hear a great many objections raised to the work of the Science and Art Department. If you will allow me to say so, my connection with that department—which, I am

happy to say, remains, and which I am very proud
of—is purely honorary; and, if it appeared to me
to be right to criticise that department with mer-
ciless severity, the Lord President, if he were inclined to resent my proceedings, could do nothing
more than dismiss me. Therefore you may believe
that I speak with absolute impartiality. My impression is this, not that it is faultless, nor that it
has not various defects, nor that there are not
sundry *lacunæ* which want filling up; but that, if
we consider the conditions under which the department works, we shall see that certain defects are
inseparable from those conditions. People talk of
the want of flexibility of the Department, of its
being bound by strict rules. Now, will any man
of common sense who has had anything to do with
the administration of public funds or knows the
humour of the House of Commons on these matters—will any man who is in the smallest degree
acquainted with the practical working of State
departments of any kind, imagine that such a
department could be other than bound by minutely
defined regulations? Can he imagine that the
work of the department should go on fairly and in
such a manner as to be free from just criticism,
unless it were bound by certain definite and fixed
rules? I cannot imagine it.

The next objection of importance that I have
heard commonly repeated is that the teaching is
too theoretical, that there is insufficient practical
teaching. I venture to say that there is no one

who has taken more pains to insist upon the comparative uselessness of scientific teaching without practical work than I have; I venture to say that there are no persons who are more cognisant of these defects in the work of the Science and Art Department than those who administer it. But those who talk in this way should acquaint themselves with the fact that proper practical instruction is a matter of no small difficulty in the present scarcity of properly taught teachers, that it is very costly, and that, in some branches of science, there are other difficulties which I won't allude to. But it is a matter of fact that, wherever it has been possible, practical teaching has been introduced, and has been made an essential element in examination; and no doubt if the House of Commons would grant unlimited means, and if proper teachers were to hand, as thick as blackberries, there would not be much difficulty in organising a complete system of practical instruction and examination ancillary to the present science classes. Those who quarrel with the present state of affairs would be better advised if, instead of groaning over the shortcomings of the present system, they would put before themselves these two questions—Is it possible under the conditions to invent any better system? Is it possible under the conditions to enlarge the work of practical teaching and practical examination which is the one desire of those who administer the department? That is all I have to say upon that subject,

Supposing we have this teaching of what I may call intermediate science, what we want next is technical instruction, in the strict sense of the word technical; I mean instruction in that kind of knowledge which is essential to the successful prosecution of the several branches of trade and industry. Now, the best way of obtaining this end is a matter about which the most experienced persons entertain very diverse opinions. I do not for one moment pretend to dogmatise about it; I can only tell you what the opinion is that I have formed from hearing the views of those who are certainly best qualified to judge, from those who have tested the various methods of conveying this instruction. I think we have before us three possibilities. We have, in the first place, trade schools—I mean schools in which branches of trade are taught. We have, in the next place, schools attached to factories for the purpose of instructing young apprentices and others who go there, and who aim at becoming intelligent workmen and capable foremen. We have, lastly, the system of day classes and evening classes. With regard to the first there is this objection, that they can be attended only by those who are not obliged to earn their bread, and consequently that they will reach only a very small fraction of the population. Moreover, the expense of trade schools is enormous, and those who are best able to judge assure me that, inasmuch as the work which they do is carrying out undertakings of this kind, which at first, at any rate, must be to a great extent tentative and experimental, by private effort. I don't believe that the man lives at this present time who is competent to organise a final system of technical education. I believe that all attempts made in that direction must for many years to come be experimental, and that we must get to success through a series of blunders. Now that work is far better performed by private enterprise than in any other way. But there is another method which I think is permissible, and not only permissible but highly recommendable in this case, and that is the method of allowing the locality itself in which any branch of industry is pursued to be its own judge of its own wants, and to tax itself under certain conditions for the purpose of carrying out any scheme of technical education adapted to its needs. I am aware that there are many extreme theorists of the individualist school who hold that all this is very wicked and very wrong, and that by leaving things to themselves they will get right. Well, my experience of the world is that things left to themselves don't get right. I believe it to be sound doctrine that a municipality— and the State itself for that matter—is a corporation existing for the benefit of its members, and that here, as in all other cases, it is for the majority to determine that which is for the good of the whole, and to act upon that. That is the principle

which underlies the whole theory of government
in this country, and if it is wrong we shall have
to go back a long way. But you may ask me,
" This process of local taxation can only be carried
out under the authority of an Act of Parliament,
and do you propose to let any municipality or any
local authority have *carte blanche* in these matters ;
is the Legislature to allow it to tax the whole
body of its members to any extent it pleases and
for any purposes it pleases ? " I should reply,
certainly not.

Let me point out to you that at this present
moment it passes the wit of man, so far as I know,
to give a legal definition of technical education.
If you expect to have an Act of Parliament with a
definition which shall include all that ought to be
included, and exclude all that ought to be excluded,
I think you will have to wait a very long time. I
imagine the whole matter is in a tentative state.
You don't know what you will be called upon to
do, and so you must try and you must blunder.
Under these circumstances it is obvious that there
are two alternatives. One of these is to give a free
hand to each locality. Well, it is within my knowledge that there are a good many people with
wonderful, strange, and wild notions as to what
ought to be done in technical education, and it is
quite possible that in some places, and especially
in small places, where there are few persons who
take an interest in these things, you will have

very remarkable projects put forth, and in that case the sole court of appeal for those taxpayers, who did not approve of such projects, would be a court of law. I suppose the judges would have to settle what is technical education. That would not be an edifying process, I think, and certainly it would be a very costly one. The other alternative is the principle adopted in the bill of last year now abandoned. I don't say whether the bill was right or wrong in detail. I am dealing now only with the principle of the bill, which appears to me to have been very often misunderstood. It has been said that it gave the whole of technical education into the hands of the Science and Art Department. It appears to me nothing could be more unfounded than that assertion. All I understand the Government proposed to do was to provide some authority who should have power to say in case any scheme was proposed, "Well, this comes within the four corners of the Act of Parliament, work it as you like;" or if it was an obviously questionable project, should take upon itself the responsibility of saying, "No, that is not what the Legislature intended; amend your scheme." There was no initiative, no control; there was simply this power of giving authority to decide upon the meaning of the Act of Parliament to a particular department of the State, whichever it might be; and it seems to me that that is a very much simpler and better process than relegating the whole question to the

of a big bucket ; and passing him afterwards
through the training college, where his life is
devoted to filling the bucket from the pump from
morning till night, without time for thought or
reflection, is a system which should not continue.
Let me assure you that it will not do for us, that
you had better give the attempt up than try that
system. I remember somewhere reading of an
interview between the poet Southey and a good
Quaker. Southey was a man of marvellous powers
of work. He had a habit of dividing his time
into little parts each of which was filled up, and
he told the Quaker what he did in this hour and
that, and so on through the day until far into the
night. The Quaker listened, and at the close said,
" Well, but, friend Southey, when dost thee
think ? " The system which I am now adverting
to is arraigned and condemned by putting that
question to it. When does the unhappy pupil
teacher, or over-drilled student of a training
college, find any time to think ? I am sure if I were
in their place I could not. I repeat, that kind of
thing will not do for science teachers. For science
teachers must have knowledge, and knowledge is
not to be acquired on these terms. The power of
repetition is, but that is not knowledge. The
knowledge which is absolutely requisite in dealing
with young children is the knowledge you
possess, as you would know your own business,
and which you can just turn about as if you

were explaining to a boy a matter of everyday life.

So far as science teaching and technical education are concerned, the most important of all things is to provide the machinery for training proper teachers. The Department of Science and Art has been at that work for years and years, and though unable under present conditions to do so much as could be wished, it has, I believe, already begun to leaven the lump to a very considerable extent. If technical education is to be carried out on the scale at present contemplated, this particular necessity must be specially and most seriously provided for. And there is another difficulty, namely, that when you have got your science or technical teacher it may not be easy to keep him. You have educated a man—a clever fellow very likely—on the understanding that he is to be a teacher. But the business of teaching is not a very lucrative and not a very attractive one, and an able man who has had a good training is under extreme temptations to carry his knowledge and his skill to a better market, in which case you have had all your trouble for nothing. It has often occurred to me that probably nothing would be of more service in this matter than the creation of a number of not very large bursaries or exhibitions, to be gained by persons nominated by the authorities of the various science colleges and schools of the country—persons such as they thought to be well qualified for the teaching business—and to be held for a certain term of years, during which the holders should be bound to teach. I believe that some measure of this kind would do more to secure a good supply of teachers than anything else. Pray note that I do not suggest that you should try to get hold of good teachers by competitive examination. That is not the best way of getting men of that special quali-fication. An effectual method would be to ask professors and teachers of any institution to re-commend men who, to their own knowledge, are worthy of such support, and are likely to turn it to good account.

I trust I am not detaining you too long; but there remains yet one other matter which I think is of profound importance, perhaps of more import-ance than all the rest, on which I earnestly beg to be permitted to say some few words. It is the need, while doing all these things, of keeping an eye, and an anxious eye, upon those measures which are necessary for the preservation of that stable and sound condition of the whole social organism which is the essential condition of real progress, and a chief end of all education. You will all recol-lect that some time ago there was a scandal and a great outcry about certain cutlasses and bayonets which had been supplied to our troops and sailors. These warlike implements were polished as bright as rubbing could make them; they were very well sharpened; they looked lovely. But when they were applied to the test of the work of war they broke and they bent, and proved more likely to hurt the hand of him that used them than to do any harm to the enemy. Let me apply that analogy to the effect of education, which is a sharpening and polishing of the mind. You may develop the intellectual side of people as far as you like, and you may confer upon them all the skill that training and instruction can give; but, if there is not, underneath all that outside form and superficial polish, the firm fibre of healthy manhood and earnest desire to do well, your labour is absolutely in vain.

Let me further call your attention to the fact that the terrible battle of competition between the different nations of the world is no transitory phenomenon, and does not depend upon this or that fluctuation of the market, or upon any condition that is likely to pass away. It is the inevitable result of that which takes place throughout nature and affects man's part of nature as much as any other—namely, the struggle for existence, arising out of the constant tendency of all creatures in the animated world to multiply indefinitely. It is that, if you look at it, which is at the bottom of all the great movements of history. It is that inherent tendency of the social organism to generate the causes of its own destruction, never yet counter-acted, which has been at the bottom of half the catastrophes which have ruined States. We are at present in the swim of one of those vast movements in which, with a population far in excess of that which we can feed, we are saved from a catastrophe, through the impossibility of feeding them, solely by our possession of a fair share of the markets of the world. And in order that that fair share may be retained, it is absolutely necessary that we should be able to produce commodities which we can exchange with food-growing people, and which they will take, rather than those of our rivals, on the ground of their greater cheapness or of their greater excellence. That is the whole story. And our course, let me say, is not actuated by mere motives of ambition or by mere motives of greed. Those doubtless are visible enough on the surface of these great movements, but the movements themselves have far deeper sources. If there were no such things as ambition and greed in this world, the struggle for existence would arise from the same causes.

Our sole chance of succeeding in a competition, which must constantly become more and more severe, is that our people shall not only have the knowledge and the skill which are required, but that they shall have the will and the energy and the honesty, without which neither knowledge nor skill can be of any permanent avail. This is what I mean by a stable social condition, because any other condition than this, any social condition in which the development of wealth involves the misery, the physical weakness, and the degradation of the worker, is absolutely and infallibly doomed to collapse. Your bayonets and cutlasses will break under your hand, and there will go on accumulating in society a mass of hopeless, physically incompetent, and morally degraded people, who are, as it were, a sort of dynamite which, sooner or later, when its accumulation becomes sufficient and its tension intolerable, will burst the whole fabric.

I am quite aware that the problem which I have put before you and which you know as much about as I do, and a great deal more probably, is one extremely difficult to solve. I am fully aware that one great factor in industrial success is reasonable cheapness of labour. That has been pointed out over and over again, and is in itself an axiomatic proposition. And it seems to me that of all the social questions which face us at this present time, the most serious is how to steer a clear course between the two horns of an obvious dilemma. One of these is the constant tendency of competition to lower wages beyond a point at which man can remain man—below a point at which decency and cleanliness and order and habits of morality and justice can reasonably be expected to exist. And the other horn of the dilemma is the difficulty of maintaining wages above this point consistently with success in industrial competition. I have not the remotest conception how this problem will eventually work itself out; but of this I am perfectly convinced, that the sole course compatible with safety lies between the two extremes ; between the Scylla of successful industrial production with a degraded population, on the one side, and the Charybdis of a population, maintained in a reasonable and decent state, with failure in industrial competition, on the other side. Having this strong conviction, which, indeed, I imagine must be that of every person who has ever thought seriously about these great problems, I have ventured to put it before you in this bare and almost cynical fashion because it will justify the strong appeal, which I make to all concerned in this work of promoting industrial education, to have a care, at the same time, that the conditions of industrial life remain those in which the physical energies of the population may be maintained at a proper level ; in which their moral state may be cared for ; in which there may be some rays of hope and pleasure in their lives; and in which the sole prospect of a life of labour may not be an old age of penury.

These are the chief suggestions I have to offer to you, though I have omitted much that I should like to have said, had time permitted. It may be that some of you feel inclined to look upon them as the Utopian dreams of a student. If there be

such, let me tell you that there are, to my
knowledge, manufacturing towns in this country,
not one-tenth the size, or boasting one-hundredth
part of the wealth, of Manchester, in which I do
not say that the programme that I have put before
you is completely carried out, but in which, at any
rate, a wise and intelligent effort had been made
to realise it, and in which the main parts of the
programme are in course of being worked out.
This is not the first time that I have had the
privilege and pleasure of addressing a Manchester
audience. I have often enough, before now, thrown
myself with entire confidence upon the hard-
headed intelligence and the very soft-hearted
kindness of Manchester people, when I have had
a difficult and complicated scientific argument to
put before them. If, after the considerations which
I have put before you—and which, pray be it
understood, I by no means claim particularly for
myself, for I presume they must be in the minds
of a large number of people who have thought
about this matter—if it be that these ideas commend themselves to your mature reflection, then
I am perfectly certain that my appeal to you to
carry them into practice, with that abundant
energy and will which have led you to take a foremost part in the great social movements of our
country many a time beforehand, will not be made
in vain. I therefore confidently appeal to you to
let those impulses once more have full sway, and

not to rest until you have done something better and greater than has yet been done in this country in the direction in which we are now going. I heartily thank you for the attention which you have been kind enough to bestow upon me. The practice of public speaking is one I must soon think of leaving off, and I count it a special and peculiar honour to have had the opportunity of speaking to you on this subject to-day.

THE END OF VOL. III

RICHARD CLAY AND SONS, LIMITED, LONDON AND BUNGAY.